Handbook

September 1994

British Bus Publishing

The North West Bus Handbook

The North West Bus Handbook is part of the Bus Handbook series that details the fleets of stage carriage and express coach operators. Where space allows other significant operators in the areas covered are also included. These handbooks are published by *British Bus Publishing* and cover Scotland, Wales and England north of London. The current list is shown at the end of the book. Together with similar books for southern England published by Capital Transport, they provide comprehensive coverage of all the principal operators' fleets in the British Isles. Handbooks for Badgerline Group and Stagecoach are also published annually.

The operators included in this edition cover those who provide stage and express services in the counties of Cumbria, Greater Manchester, Lancashire and Merseyside. Also included are a number of those operators who provide significant coaching activities.

Quality photographs for inclusion in these, and other areas covered by the series are welcome, for which a fee is payable. The publishers unfortunately cannot accept responsibility for any loss and request you clearly show your initials on each picture or slide. Details of changes to fleet information are also welcome.

More information on the Bus Handbook series is available from:

British Bus Publishing,
The Vyne,
16 St Margaret's Drive
Wellington
Telford,
Shropshire TF1 3PH

Series Editor: Bill Potter

Principal Editors for *The North West Bus Handbook*:
Graham Ashworth, David Donati, and Simon Watts

Acknowledgements:
We are grateful to Paul Clayton, Keith Grimes, Mark Jameson, Colin Lloyd, Brian Pritchard, Capital Transport, the PSV Circle and the operating companies for their assistance in the compilation of this book.

The front cover photo is by Paul Wigan
The rear cover photographs are by Steve Sanderson and Paul Wigan

Contents correct to September 1994

ISBN 1 890990 07 3
Published by *British Bus Publishing*
The Vyne, 16 St Margarets Drive, Wellington,
Telford, Shropshire, TF1 3PH

R. BULLOCK
R. BULLOCK
42 PICCADILLY
GM buses
371
Danger
Look left
L20 BUL

PIER HEAD 212
LIVERBUS
Liverbus
L211 SBG
TRUCTI

Contents

Overleaf, top: **The Bullock fleet has recently been bolstered by a pair of new Leyland Olympians with East Lancashire bodywork. Photographed as it crosses the tram lines in Piccadilly, Manchester is L20BUL.** *Keith Grimes*

Overleaf, bottom: **Eleven Volvo Olympians joined the Gemsam fleet in 1993, ten dual-doored versions being allocated to the London Suburban operation based in Edmonton, while the eleventh, 211, L211SBG, has a single doorway and is lettered for Liverbus work in Liverpool.** *Phillip Stephenson*

A1A TRAVEL

J Ashworth, 158-164 Price Street, Birkenhead, Merseyside, L41 3PR

1	L1BUS	Mercedes-Benz 609D	North West CS	B17F	1993	
2	L2BUS	Mercedes-Benz 609D	North West CS	B17F	1993	
3	A3JUG	Mercedes-Benz L608D	PMT Hanbridge	C21F	1985	Ex Acorn, York, 1991
5	SXI9035	Mercedes-Benz L207D	Devon Conversions	M12	1980	Ex BRC Tockington, 1990
6	SXI9036	Mercedes-Benz L608D	Reeve Burgess	C21F	1984	Ex Wirral Christian Centre, 1991
7	E127RAX	Freight Rover Sherpa	Carlyle Citybus 2	B20F	1987	Ex Phil Anslow, Garndiffaith, 1993
8	F452XON	Freight Rover Sherpa	Carlyle Citybus 2	B20F	1988	Ex Midland, 1994
9	D579EWS	Freight Rover Sherpa	Dormobile	B16F	1986	Ex The Wright Company, Wrexham, 1994
10	E933RWR	Freight Rover Sherpa	Carlyle Citybus 2	B18F	1987	Ex Chesterfield Omnibus, 1994
11	E936RWR	Freight Rover Sherpa	Carlyle Citybus 2	B18F	1987	Ex Chesterfield Omnibus, 1994
12	H62WNN	Mercedes-Benz 709D	Scott	C29F	1991	Ex Skills, Nottingham, 1992
15	AIA5505	Volkswagen LT55	Optare City Pacer	B25F	1986	Ex Torkard, Hucknall, 1994
16	D362JUM	Volkswagen LT55	Optare City Pacer	B25F	1987	Ex Aston Express, 1993
17	OIA773	Bedford CFL	Plaxton Mini Supreme	C17F	1980	Ex Ruxley, Tolworth, 1989
18	OED3W	Mercedes-Benz L508D	Devon Conversions	C18F	1981	Ex Whites of Calver, 1993
19	E319LHG	Peugeot-Talbot Pullman	Talbot	B22F	1988	Ex Pedley, Hammerwick, 1993
20	AIA1120	Volkswagen LT55	Optare City Pacer	B25F	1988	Ex Phil Anslow, Garndiffaith, 1993
21	F430BFY	Mazda E2200	Made-to-Measure	M14	1989	
22	F212AKG	Freight Rover Sherpa	Carlyle Citybus 2	B20F	1988	Ex Shamrock, Pontypridd, 1993

Previous Registrations:

A3JUG	B852ETU, AIA9000	OIA773	KBH854V	SXI9036	B296ADM
F430BFY	F636HVM, AIA9000	SXI9035	DFB158W		

Livery: Cream and red

A1A Travel operate on the Wirral penisular, the northern part of which 'moved' from Cheshire to Merseyside in 1974. Photographed while heading for Upton is Freight Rover Sherpa E936RWR, one of the later examples from Carlyle which employs the Citybus 2 design. *Phillip Stephenson*

A2B TRAVEL

W G Evans, 5 The Mount, Heswall, Merseyside, L60 7RE

D737BDC	Mercedes-Benz L307D	Reeve Burgess	M12	1986	Ex Cooper, Gilesgate Moor, 1993
D631RTM	Mercedes-Benz L307D	Reeve Burgess	M12	1987	Ex Time Transport, Thornton Heath, 1992
E474FPN	Ford Transit VE6	Dormobile	B20F	1988	Ex Ericsson, Horsham, 1992
E56MMT	Mercedes-Benz L307D	Reeve Burgess	M12	1987	Ex Time Transport, Thornton Heath, 1992
F90JGE	Ford Transit VE6	Dormobile	B20F	1989	Ex Pathfinder, Newark, 1992
F130KAO	Mercedes-Benz 609D	Reeve Burgess	B20F	1989	Ex North Western, 1994
F136KAO	Mercedes-Benz 609D	Reeve Burgess	B20F	1989	Ex North Western, 1994
F357TSX	Mercedes-Benz 811D	Alexander AM	C33F	1988	Ex Anderson, Lower Largo, 1993
G486PNF	Mazda E2200	Made-to-Measure	M14	1990	Ex Vickers, Morecambe, 1994
G145TGX	Mercedes-Benz 811D	Phoenix	DP29F	1990	Ex Kentish Bus, 1993
A2BEO	Mercedes-Benz 609D	Made-to-Measure	C26F	1989	
H506BND	Mercedes-Benz 811D	Made-to-Measure	C19F	1990	
H794HEM	Mercedes-Benz 408D	Made-to-Measure	M15	1991	
L35AKP	Iveco TurboDaily 59.12	Dormobile Routemaker	B27F	1993	
L177PDO	Mercedes-Benz 814D	Autobus Classique	C33F	1993	

Previous Registrations:
A2BEO G639UHU

Livery: White, blue and red; black, yellow and white (Merseytravel) L35AKP.

Mercedes-Benz 811D F357TSX, seen here in Chester bus station, carries an Alexander AM coach body and now operates for A2B Travel. It joined the fleet last year from Anderson of Lower Largo.
P Stockwell

ABBOTT

J Abbott & Sons (Blackpool) Ltd, 95/97 Talbot Road, Blackpool, Lancashire, FY1 3QX

OFV467G	AEC Reliance 6U3ZR	Plaxton Elite	C57F	1969	
SFR127J	AEC Reliance 6U3ZR	Plaxton Elite	C53F	1970	
TFV147J	AEC Reliance 6U3ZR	Plaxton Elite	C53F	1971	
WFR147K	AEC Reliance 6U3ZR	Plaxton Elite II	C45F	1972	
WFR167K	AEC Reliance 6U3ZR	Plaxton Elite II	C45F	1972	
NFR487M	AEC Reliance 6U3ZR	Plaxton Elite III	C45F	1973	
NFR497M	AEC Reliance 6U3ZR	Plaxton Elite III	C53F	1973	
PBV637P	AEC Reliance 6U3ZR	Plaxton Supreme III	C53F	1976	
VFV907R	AEC Reliance 6U3ZR	Plaxton Supreme III	C53F	1977	
YRN507R	AEC Reliance 6U3ZR	Plaxton Supreme III	C53F	1977	
HCK847S	AEC Reliance 6U3ZR	Duple Dominant II	C53F	1978	
JCW517S	Volvo B58-56	Plaxton Supreme III	C53F	1978	
NCW747T	AEC Reliance 6U3ZR	Duple Dominant II	C53F	1979	
EFR97W	Leyland Leopard PSU5D/5R	Duple Dominant II	C53F	1981	
EFR107W	Leyland Leopard PSU5D/5R	Duple Dominant II	C53F	1981	
MSU611Y	Volvo B10M-61	Duple Laser	C57F	1983	Ex Duple demonstrator, 1983
A547HBV	Leyland Tiger TRCTL11/3R	Duple Caribbean	C51F	1983	
A777RBV	Leyland Tiger TRCTL11/3R	Duple Caribbean	C51F	1984	
B357UCW	Leyland Tiger TRCTL11/3R	Duple Laser	C55F	1984	Ex Duple demonstrator, 1985
D857XBV	Leyland Tiger TRCTL11/3RZ	Duple 340	C53FT	1987	
D327VVV	Quest J	Jonckheere Piccolo	C33FT	1987	
F727VAC	Volvo B10M-60	Plaxton Paramount 3500 III	C53F	1988	Ex Volvo Bus demonstrator, 1989
H687XBV	Volvo B10M-60	Plaxton Paramount 3500 III	C57F	1991	
K617SBV	Volvo B10M-60	Jonckheere Deauville	C53FT	1992	

Livery: Red and grey; WFR147K is operated by Blackhurst and YRN507R by Enterprise Motors.

Abbotts of Blackpool now occupy the Talbot Road depot previously used by Ribble. They still run many AEC Reliances, all of which were purchased new. The express service from Fleetwood to Manchester has continued for more than fifty years, at one time being a joint service with Lansdown coaches. In recent times the service has been extended at the Manchester end to Hollinwood and Oldham. Photographed in Blackpool is HCK847S.
Paul Wigan

ABC TRAVEL

D & M Garnett, 7 Rothwell Drive, Ainsdale, Southport, Merseyside, PR8 2SB

TCK823	Leyland Titan PD3/5	MCW	H41/31F	1963	Ex 1823 Preservation Group, 1994
ABC21Y	Leyland Tiger TRCTL11/3R	Duple Dominant IV	C53F	1983	Ex Smith, Amble, 1993
ABC24Y	Leyland Tiger TRCTL11/3RH	Plaxton Paramount 3200 E	C57F	1983	Ex Redline, Penwortham, 1994
ABC75Y	Leyland Olympian ONTL11/2RSp	Eastern Coach Works	CH45/28F	1983	Ex Clyde Coast, Ardrossan, 1992
D589MVR	Leyland Tiger TRCTL11/3RZ	Plaxton Paramount 3200 III	C53F	1987	Ex Shearings, 1992
F352DVR	Ford Transit VE6	Mellor	B16F	1989	Ex Gascoine, Partridge Green, 1990
F354DVR	Ford Transit VE6	Mellor	B16F	1989	Ex Gascoine, Partridge Green, 1990
F203XBV	Freight Rover Sherpa	Carlyle Citybus 2	B20F	1989	
F226AWO	Freight Rover Sherpa	Carlyle Citybus 2	B20F	1989	Ex National Welsh, 1992
F777GNA	Leyland Tiger TRCTL11/3RZ	Plaxton Paramount 3200 III	C53F	1989	Ex Shearings, 1993
H4ABC	Optare MetroRider	Optare	B33F	1990	
J55ABC	Mercedes-Benz 709D	Alexander AM	DP25F	1991	
J555ABC	Mercedes-Benz 709D	Dormobile Routemaker	B29F	1992	
J44ABC	Optare MetroRider	Optare	B33F	1992	
J444ABC	Optare MetroRider	Optare	B33F	1992	
J800ABC	DAF SB220LC550	Optare Delta	B49F	1992	Ex Optare demonstrator, 1994
K400ABC	Optare MetroRider	Optare	B33F	1992	
L700ABC	DAF SB220LC550	Optare Delta	B49F	1994	
L5ABC	Optare MetroRider	Optare	B23F	1994	
L6ABC	Optare MetroRider	Optare	B23F	1994	
L99ABC	Mercedes-Benz 709D	Marshall C19	B23F	1994	
L999ABC	Mercedes-Benz 709D	Marshall C19	B23F	1994	

Previous Registrations:

ABC21Y	YPD117Y	ABC75Y	YPJ503Y, 341AYF, HSB312Y
ABC24Y	A214DPB	J800ABC	J365BNW

Livery: Red and cream.

Seen in Liverpool on a Merseytravel service is ABC Travel J555ABC, the sole Mercedes-Benz 709D with Dormobile Routemaker bodywork in the fleet. ABC Travel have recently purchased two DAF SB220LC550s with Optare Delta bodywork for their service to Skipton, one of which can be seen on the back cover of this Bus Handbook.
Roy Marshall

AINTREE COACHLINE

J Cherry, 11 Clare Road, Bootle, Merseyside, L20 9LY

A1	ANE2T	Leyland Titan TNLXB/1RF	Park Royal	H47/26F	1979	Ex The Wright Company, Wrexham, 1991
A2	JPT905T	Bristol VRT/SL3/501	Eastern Coach Works	DPH39/31F	1979	Ex Go-Ahead, 1993
A3	CUL72V	Leyland Titan TNLXB/2RRSp	Park Royal	H44/34F	1980	Ex London Buses, 1992
A4	TUB250R	Foden-NC/6LXB	Northern Counties	H47/26D	1977	Ex WMPTE, 1989
A5	AFY185X	Leyland Atlantean AN68B/1R	Willowbrook	H45/33F	1981	Ex Somerbus, Paulton, 1994
A6	RTJ423L	Daimler Fleetline CRG6LXB	Northern Counties	H49/30F	1972	Ex Platt Bridge Morris Dancers, 1994
A7	GDB164N	Daimler Fleetline CRG6LXB	Northern Counties	H49/30F	1974	Ex Lochview, Greenock, 1994
L1	DCM552X	Leyland Tiger TRCTL11/3R	Duple Dominant II	C53F	1983	Ex Somerbus, Paulton, 1994
L2	D561HNW	Iveco Daily 49.10	Robin Hood City Nippy	DP21F	1987	Ex Bu-Val, Littleborough, 1994
L5	831HKA	Mercedes-Benz 0303	Mercedes-Benz	C49FT	1985	Ex Village, Garston, 1993
L6	LIB4057	Volvo B58-61	Jonckheere Bermuda	C51FT	1982	Ex Arena Travel, Speke, 1991
M2	A507GPC	Leyland Olympian ONTL11/2R	Eastern Coach Works	CH45/28F	1984	Ex Auty, Rushden, 1992

Previous Registrations:
831HKA B146MJV DCM552X WVR60X, HCK466 LIB4057 WRA28Y

Livery: Blue and white

ATHERTON BUS COMPANY

P G Hughes, 33 Landedmans, Westhoughton, Bolton, Greater Manchester, BL5 2QJ

CNH171X	Leyland Leopard PSU3F/4R	Eastern Coach Works B51	C49F	1981	Ex United Counties, 1994

Livery: Blue, white and yellow.

Atherton Bus Company have recently replaced their double deck buses with a Leyland Leopard. Seen working service 592 from Bolton's Moor Lane bus station is CNH171X an Eastern Coach Works-bodied example recently acquired from United Counties.
Andy Jarosz

AVON BUSES

L W Smith, Unit 1, Tarran Industrial Estate, Moreton, Wirral, Merseyside, L46 4TP

AV60	ANC578A	Ailsa B55-10	Alexander AV	H44/35F	1975	Ex London Buses, 1991
	KSD102W	Volvo B55-10	Alexander AV	H44/35F	1980	Ex Clydeside 2000, 1993
LN74	C974PFS	Leyland National 2 NL116L11/1R		B52F	1986	Ex Scottish Prisons, 1992
NG63	WUH163T	Leyland National 11351A/1R	East Lancs Greenway (1993)	B48F	1978	Ex Rhondda, 1993
T68	CUL68V	Leyland Titan TNLXB/2RRSp	Park Royal	H44/26D	1980	Ex London Buses, 1992
T96	CUL96V	Leyland Titan TNLXB/2RRSp	Park Royal	H44/26D	1980	Ex London Buses, 1992

Livery: Blue and cream

Previous Registrations:

ANC578A JOV760P

B & D COACHES

R Dootson, 12 Northolt Avenue, Leigh, Greater Manchester, WN7 1TY

D23KAX	Iveco Daily 49.10	Robin Hood City Nippy	B21F	1986	Ex Cynon Valley, 1992
D502MJA	Iveco Daily 49.10	Robin Hood City Nippy	B19F	1986	Ex Midland Choice, Codnor, 1992
D511MJA	Iveco Daily 49.10	Robin Hood City Nippy	B19F	1986	Ex Midland Choice, Codnor, 1992
D520MJA	Iveco Daily 49.10	Robin Hood City Nippy	B19F	1986	Ex GM Buses, 1992
D615BCK	Iveco Daily 49.10	Robin Hood	B21F	1987	Ex Ribble, 1993
D178LTA	Renault-Dodge S56	Reeve Burgess	B23F	1986	Ex Cardiff, 1994
D181LTA	Renault-Dodge S56	Reeve Burgess	B23F	1986	Ex Cardiff, 1994
F248KVU	Renault-Dodge S46	Northern Counties	B24F	1989	Ex City Nippy, Middleton, 1992
F249KVU	Renault-Dodge S46	Northern Counties	B24F	1989	Ex City Nippy, Middleton, 1992
F220AKG	Iveco Daily 49.10	Carlyle Dailybus 2	B21F	1988	Ex Cynon Valley, 1992

Livery: Blue and grey.

The name B & D Coaches comes from the original partners, Bellair and Dootson who have developed a network of minicoaches in the Atherleigh area of the Wigan District in Greater Manchester. Photographed departing from Leigh bus station is F248KVU, a Northern Counties-bodied Renault-Dodge S46 previously found running to the north of Manchester with City Nippy.
Roy Marshall

BLACKBURN

Blackburn Borough Transport Ltd, Intack Garage, Whitebirk Road,
Blackburn, Lancashire, BB1 3JD

6-22 Leyland Atlantean AN68C/1R East Lancashire H43/31F 1982

6	OCW6X	**10**	OCW10X	**14**	OCW14X	**17**	SBV17X	**20**	VBV20Y
7	OCW7X	**11**	OCW11X	**15**	OCW15X	**18**	VBV18Y	**21**	VBV21Y
8	OCW8X	**12**	OCW12X	**16**	SBV16X	**19**	VBV19Y	**22**	VBV22Y
9	OCW9X	**13**	OCW13X						

23-29 Leyland Atlantean AN68D/1R East Lancashire H43/31F 1983

23	A23JBV	**25**	FCK25Y	**27**	FCK27Y	**28**	A28JBV	**29**	A29JBV
24	FCK24Y	**26**	A26JBV						

43	ABV43B	Leyland Titan PD2A/24	East Lancashire	H35/28R	1964
115	PCW115P	Leyland Atlantean AN68A/1R	East Lancashire	H45/31F	1976

117-124 Leyland Atlantean AN68A/1R East Lancashire H45/31F 1977-78

117	CBV117S	**119**	CBV119S	**121**	CBV121S	**123**	CBV123S	**124**	CBV124S
118	CBV118S	**120**	CBV120S	**122**	CBV122S				

The traditional combination of Leyland/East Lancashire can still be found in many of the former Lancashire fleets, Blackburn being no exception. Photographed on Blackburn bus station, and destined for Accrington, is 18, VBV18Y. *Phillip Stephenson*

Five examples of the Volvo B10M in the Blackburn fleet are fitted with Caetano Algarve bodywork, the vehicles being used on an inclusive holiday programme, including several tours onto continental Europe. However 318, PCB24, was seen at Staples Corner while heading for the less romantic Wembley in May 1994. *Keith Grimes*

125-142 Leyland Atlantean AN68A/1R — East Lancashire — H43/31F — 1979-80

125	LFR125T	**129**	LFR129T	**135**	WRN135V	**138**	WRN138V	**141**	WRN141V
126	LFR126T	**130**	LFR130T	**136**	WRN136V	**139**	WRN139V	**142**	WRN142V
127	LFR127T	**133**	WRN133V	**137**	WRN137V	**140**	WRN140V		
128	LFR128T	**134**	WRN134V						

143-147 Leyland Atlantean AN68C/1R — East Lancashire — H43/31F — 1981

143	DBV143W	**144**	DBV144W	**145**	DBV145W	**146**	DBV146W	**147**	DBV147W

201-205 Volvo B10M-55 — East Lancashire EL2000 — DP51F — 1991

201	J421JBV	**202**	J422JBV	**203**	J418JBV	**204**	J419JBV	**205**	J420JBV

303	KJF3V	Leyland Leopard PSU3E/4R	Plaxton Supreme IV Express	C53F	1980	Ex Leicester, 1986
304	KBC4V	Leyland Leopard PSU3E/4R	Plaxton Supreme IV Express	C53F	1980	Ex Leicester, 1986
305	FUT5V	Leyland Leopard PSU3E/4R	Plaxton Supreme IV Express	C53F	1979	Ex Leicester, 1986
306	FUT6V	Leyland Leopard PSU3E/4R	Plaxton Supreme IV Express	C53F	1979	Ex Leicester, 1986
307	LHE254W	Leyland Leopard PSU5E/4R	Plaxton Supreme IV	C53F	1981	Ex National Travel East, 1987
308	LHE255W	Leyland Leopard PSU5E/4R	Plaxton Supreme IV	C53F	1981	Ex National Travel East, 1987
317	B417CMC	Leyland Tiger TRCTL11/3R	Plaxton Paramount 3200	C53F	1985	Ex British Airways, 1987
318	PCB24	Volvo B10M-61	Caetano Algarve	C49FT	1986	Ex Park's, Hamilton, 1988
319	D319NEC	Leyland Tiger TRCTL11/3RH	Plaxton Paramount 3200 III	C53F	1987	
320	D320NEC	Leyland Tiger TRCTL11/3RH	Plaxton Paramount 3200 III	C53F	1987	
321	F984HGE	Volvo B10M-60	Plaxton Paramount 3500 III	C53F	1988	Ex Park's, Hamilton, 1991
322	MIB920	Volvo B10M-61	Caetano Algarve	C49FT	1987	Ex The King's Ferry, Gillingham, 1991
323	G996RKN	Volvo B10M-60	Caetano Algarve	C49FT	1990	Ex The King's Ferry, Gillingham, 1991
324	LIJ749	Volvo B10M-61	Caetano Algarve	C49FT	1988	Ex The King's Ferry, Gillingham, 1992
325	G999RKN	Volvo B10M-60	Caetano Algarve	C49FT	1990	Ex The King's Ferry, Gillingham, 1992
326	H174EJU	Volvo B10M-60	Plaxton Paramount 3500 III	C51F	1991	Ex Ambassador Travel, 1994
410	PUS158W	Leyland Leopard PSU3F/4R	Alexander AYS	B53F	1981	Ex Kelvin Scottish, 1987
411	TSU639W	Leyland Leopard PSU3G/4R	Alexander AYS	B53F	1981	Ex Kelvin Scottish, 1987
412	TSU640W	Leyland Leopard PSU3G/4R	Alexander AYS	B53F	1981	Ex Kelvin Scottish, 1987
413	TSU641W	Leyland Leopard PSU3G/4R	Alexander AYS	B53F	1981	Ex Kelvin Scottish, 1988
414	WPH134Y	Leyland Tiger TRCTL11/2R	Eastern Coach Works B51	DP51F	1982	Ex Kentish Bus, 1988
415	WPH137Y	Leyland Tiger TRCTL11/2R	Eastern Coach Works B51	DP51F	1982	Ex Kentish Bus, 1988
416	WPH138Y	Leyland Tiger TRCTL11/2R	Eastern Coach Works B51	DP51F	1982	Ex Kentish Bus, 1988
417	WPH141Y	Leyland Tiger TRCTL11/2R	Eastern Coach Works B51	DP51F	1982	Ex Kentish Bus, 1988
423	BVP807V	Leyland National 2 NL116L11/1R	East Lancs Greenway (1993)	B49F	1980	Ex North Western, 1992
424	AFM5W	Leyland National 2 NL116AL11/2R	East Lancs Greenway (1993)	B49F	1981	Ex North Western, 1992
425	PJT267R	Leyland National 11351A/2R	East Lancs Greenway (1994)	B49F	1976	Ex Solent Blue Line, 1994
426	TCW868T	Leyland National 11351A/2R	East Lancs Greenway (1994)	B49F	1979	Ex North Devon, 1994
525	BFV861R	Leyland National 10351A/2R	East Lancs Greenway (1993)	B41F	1977	Ex London Buses, 1992
526	BYW361V	Leyland National 10351A/2R	East Lancs Greenway (1993)	B41F	1979	Ex Evag Cannon, Bolton, 1992
527	LRN552N	Leyland National 10351/2R	East Lancs Greenway (1993)	B41F	1975	Ex Isle of Man, 1992
528	JDT432N	Leyland National 10351/2R	East Lancs Greenway (1993)	B41F	1975	Ex Isle of Man, 1992
529	JWG191P	Leyland National 10351/2R	East Lancs Greenway (1993)	B41F	1975	Ex Isle of Man, 1992
530	HPF313N	Leyland National 10351/1R/SC	East Lancs Greenway (1993)	B41F	1975	Ex Birmingham Omnibus, Smethwick, 1992
531	AYR354T	Leyland National 10351A/2R	East Lancs Greenway (1994)	B41F	1979	Ex PMT, 1994

Blackburn have played a keen role in the East Lancashire Greenway project with several now operating as part of the fleet. Recently the shorter 10.3 metre examples have been renumbered into the 500 series where they have been joined by four new Volvo B6s. Photographed while showing fleet number 427 is 528, LRN552N. *Phillip Stephenson*

532	JJG907P	Leyland National 10351/1R	East Lancs Greenway (1994)	B41F	1975	Ex Busylink, Hemel Hempstead, 1994
533	M533RCW	Volvo B6-50	East Lancashire	B41F	1994	
534	M534RCW	Volvo B6-50	East Lancashire	B41F	1994	
535	M535RCW	Volvo B6-50	East Lancashire	B41F	1994	
536	M536RCW	Volvo B6-50	East Lancashire	B41F	1994	

601-611 MCW Metrorider MF151/4 — MCW — B23F* — 1987 — *603/10/1 are DP23F and M151/5

601	D601AFR	**604**	D604AFR	**606**	D606AFR	**608**	D608AFR	**610**	D610AFR
602	D602AFR	**605**	D605AFR	**607**	D607AFR	**609**	D609AFR	**611**	D611AFR
603	D603AFR								

612	E612FRN	MCW Metrorider MF150/42	MCW	B23F	1987
614	E614FRN	MCW Metrorider MF150/42	MCW	B23F	1987
615	E615FRN	MCW Metrorider MF150/42	MCW	B23F	1987

616-626 MCW Metrorider MF159/3 — MCW — B33F* — 1988 — *625/6 are DP33F and MF159/4

616	F616UBV	**619**	F619UBV	**621**	F621UBV	**623**	F623UBV	**625**	F625UBV
617	F617UBV	**620**	F620UBV	**622**	F622UBV	**624**	F624UBV	**626**	F626UBV
618	F617UBV								

Previous Registrations:

BFV861R	OJD880R	LRN552N	JDT437N, 3176MAN
JDT432N	JDT432N, 4647MAN	MIB920	E841EUT
JWG191P	JWG191P, 4648MAN	PCB24	C673KDS
LIJ749	F870TNH	TCW868T	AFJ757T

Livery: Ivory and green

BLACKPOOL

Blackpool Transport Services Ltd, Rigby Road, Blackpool,
Lancashire, FY1 5DD

101-130 DAF SB220LC550 | Optare Delta | DP46F | 1990-93

101	G101NBV	**107**	G107NBV	**114**	H114YHG	**120**	H120CHG	**126**	J126GRN
102	G102NBV	**108**	G108NBV	**115**	H115YHG	**121**	H121CHG	**127**	K127UFV
103	G103NBV	**109**	H109YHG	**116**	H116YHG	**122**	H122CHG	**128**	K128UFV
104	G104NBV	**110**	H110YHG	**117**	H117YHG	**123**	J123GRN	**129**	K129UFV
105	G105NBV	**112**	H112YHG	**118**	H118CHG	**124**	J124GRN	**130**	K130UFV
106	G106NBV	**113**	H113YHG	**119**	H119CHG	**125**	J125GRN		

322-340 Leyland Atlantean AN68A/2R | East Lancashire | H50/36F | 1979-80

322	URN322V	**326**	URN326V	**330**	URN330V	**334**	AHG334V	**338**	AHG338V
323	URN323V	**327**	URN327V	**331**	AHG331V	**336**	AHG336V	**339**	AHG339V
324	URN324V	**328**	URN328V	**332**	AHG332V	**337**	AHG337V	**340**	AHG340V
325	URN325V	**329**	URN329V	**333**	AHG333V				

341-350 Leyland Atlantean AN68C/2R | East Lancashire | H50/36F | 1981

341	GHG341W	**344**	GHG344W	**346**	GHG346W	**348**	GHG348W	**350**	GHG350W
343	GHG343W	**345**	GHG345W	**347**	GHG347W	**349**	GHG349W		

351-362 Leyland Atlantean AN68D/2R | East Lancashire | H49/36F | 1982/83

351	UHG351Y	**354**	UHG354Y	**357**	A357HHG	**359**	A359HHG	**361**	A361HHG
352	UHG352Y	**355**	A355HHG	**358**	A358HHG	**360**	A360HHG	**362**	A362HHG
353	UHG353Y	**356**	A356HHG						

Opposite, top: **Blckpool was first with the electric tram and these machines are still very much a part of the Fylde coast. Five units are decorated for use during the illuminations, two coupled together to form the Wild West train, seen here during an evening run.** *Mark Bailey*

Opposite, bottom: **To augment the promenade service, twelve Routemasters are used in a red and white livery in a style similar to that used before the green scheme. Photographed near the Tower is 527, WLT879.** *Paul Wigan*

The Optare Delta is featured in both Blackpool and its new subsidiary, Blue Buses of Fylde. While the fleets continue for the present as separate units there is a probability of more interchange soon. Blackpool is the setting for this picture of 127, K127UFV, as it heads towards Grange Park on service 6. *Mike Fowler*

GREAT HOLIDAY CENTRES IN ENGLAND AND NORTH WALES....
PONTIN'S
Family Favourites
TEL. 0772 621621
40
TOWER
SANDCASTLE
PLEASURE BEACH
PONTINS
THIS WAY ENTRANCE
ACTION PACK
So many leisure facilities and no extra to pay.
CAPTAIN CROC
Come & Join THE PARTY
527
WLT 879

FISHERMAN'S FRIEND
FISHERM
FF
FISHERMAN'S FRIEND
FISHERMAN'S FRIEND
CIRCULAR TOUR

363	B363UBV	Leyland Atlantean AN68D/2R	East Lancashire	DPH45/29F	1984	
364	B364UBV	Leyland Atlantean AN68D/2R	East Lancashire	DPH45/29F	1984	
365	UWW5X	Leyland Olympian ONLXB/1R	Roe	H47/29F	1982	Ex West Yorkshire PTE, 1986
366	UWW11X	Leyland Olympian ONLXB/1R	Roe	H47/29F	1982	Ex West Yorkshire PTE, 1986
367	UWW15X	Leyland Olympian ONLXB/1R	Roe	H47/29F	1982	Ex West Yorkshire PTE, 1986

368-373 Leyland Olympian ONCL10/1RZ* East Lancashire H45/31F 1989 *372/3 are ONCL10/1RZA

368	F368AFR	**370**	F370AFR	**371**	F371AFR	**372**	F372AFR	**373**	F373AFR
369	F369AFR								

374-379 Volvo Olympian YN2RV18Z4 Northern Counties Palatine DPH45/25F 1994

374	M374	**376**	M376	**377**	M377	**378**	M378	**379**	M379
375	M375								

507	HFR507E	Leyland Titan PD3A/1	MCW	H41/30R	1967

521-531 AEC Routemaster 5RM Park Royal H36/28R 1961-64 Ex London Buses, 1986/88

521	583CLT	**524**	640DYE	**526**	735DYE	**528**	357CLT	**530**	ALD989B
522	WLT848	**525**	650DYE	**527**	WLT879	**529**	ALD966B	**531**	ALM71B
523	627DYE								

532	DHC784E	Leyland Titan PD2A/30	East Lancashire	O32/28R	1967	Ex Eastbourne, 1989
533	ALM89B	AEC Routemaster 5RM	Park Royal	H36/28R	1964	Ex London Buses, 1988

549-568 Volkswagen LT55 Optare City Pacer B21F* 1987 565/6 Ex Optare demonstrators, 1987
*563-6 are DP21F

549	E549GFR	**553**	E553GFR	**557**	E557GFR	**561**	D561YCW	**565**	D854MUA
550	E550GFR	**554**	E554GFR	**558**	E558GFR	**562**	D562YCW	**566**	D898NUA
551	E551GFR	**555**	E555GFR	**559**	E559GFR	**563**	D563YCW	**567**	E567GFR
552	E552GFR	**556**	E556GFR	**560**	D560YCW	**564**	D564YCW	**568**	E568GFR

569-582 Volkswagen LT55 Optare City Pacer B21F 1988

569	E569OCW	**572**	F572RCW	**575**	F575RCW	**578**	F578RCW	**581**	F581WCW
570	E570OCW	**573**	F573RCW	**576**	F576RCW	**579**	F579WCW	**582**	F582WCW
571	F571RCW	**574**	F574RCW	**577**	F577RCW	**580**	F580WCW		

583	F934AWW	Volkswagen LT55	Optare City Pacer	DP21F	1988	Ex Optare demonstrator, 1989

Trams

600-607 English Electric M4d English Electric OST56C* 1934-35 *602 is OST52C

600	**602**	**604**	**605**	**606**	**607**

619	English Electric M4d	Blackpool Corporation (1987)	ST52D	1973

621-637 EMB M4d Brush ST48C 1937

621	**623**	**626**	**630**	**631**	**632**	**633**	**634**	**636**	**637**
622	**625**	**627**							

641-647 Primrose/Brush M4d East Lancashire ST52D 1984-88 641 is ST55D

641	**642**	**643**	**644**	**645**	**646**	**647**

648	Maley & Taunton/Brush M4d	East Lancashire	ST53D	1985
660	Maley & Taunton M4d	Roberts	ST56C	1953

671-677 English Electric 1935 M4s English Electric ST53C 1960

671	**672**	**673**	**674**	**675**	**676**	**677**

678	English Electric 1935 M4d	English Electric	ST48C	1960
679	English Electric 1935 M4d	English Electric	ST48C	1960
680	English Electric 1935 M4d	English Electric	ST48C	1960

681-687		Maley & Taunton M4d		MCW		ST61C	1960		
681	682	683	684	685	686	687			

700-726		English Electric M4d		English Electric		DT54/40C*	1934/35	*706 is ODT54/40C	
700	703	707	710	713	716	718	720	722	724
701	704	708	711	715	717	719	721	723	726
702	706	709	712						

732	Dick Kerr M4s	Blackpool Corporation	48-seat	1960	*Moon Rocket*
733	English Electric M4d	Blackpool Corporation	35-seat	1962	*Wild West Loco*
734	Dick Kerr B4s	Blackpool Corporation	60-seat	1962	*Wild West coach*
735	English Electric M4d	Blackpool Corporation	57/42	1963	*Hovercraft*
736	Dick Kerr M4d	Blackpool Corporation	71-seat	1965	*Frigate*
761	English Electric (1934) M4d	Blackpool/East Lancs	DT56/44F	1979	
762	English Electric (1934) M4d	Blackpool/East Lancs	DT56/34D	1982	

Livery: Cream and green; red and white 507, 521/3-33; black and yellow (Handybus) 549-583

A black and yellow livery with Handybus names is used by Blackpool for its minibus operation. All the vehicles are Volkswagen LT55s with Optare City Pacer bodywork. Seen outside the tower is 578, F578RCW.

Tram 648 is the only example of the new East Lancashire-bodied fleet not to be based on Primrose trucks. Photographed in Lord Street, Fleetwood, which, until the re-introduction of trams in Manchester, was the only street in England where street trams where scheduled.
Paul Wigan

BLUEBIRD

MTG, TA & M Dunstan, Aleaxander House, Greengate, Middleton,
Greater Manchester, M24 1RY

2	K2BLU	Iveco Daily 49.10	Dormobile Routemaker	B23F	1992	
3	K3BLU	Iveco Daily 49.10	Dormobile Routemaker	B23F	1992	
4	L4BLU	Iveco Turbodaily 59.12	Marshall C31	B27F	1994	
7	E207GPV	Renault-Dodge S56	Northern Counties	B20F	1987	Ex Ipswich, 1994
8	E208GPV	Renault-Dodge S56	Northern Counties	B20F	1987	Ex Ipswich, 1994
15	YHG15V	Leyland Leopard PSU4E/4R	Alexander AY	DP45F	1980	Ex Burnley & Pendle, 1993
16	LBU607V	Leyland Leopard PSU4E/4R	Alexander AY	DP45F	1980	Ex Burnley & Pendle, 1993
17	VXP717	Leyland Tiger TRCTL11/3R	Duple Laser	C50F	1984	Ex Thames Transit, 1992
23	F223AKG	Iveco Daily 49.10	Carlyle Dailybus 2	B21F	1988	Ex Cynon Valley, 1992
28	RIB1728	Leyland Leopard PSU5C/4R	Duple Dominant II	C50F	1979	Ex Ribble, 1991
30	HSK330	Leyland Leopard PSU3E/4R	Duple Dominant II Express	C47F	1979	Ex Ribble, 1991
32	G532VND	Iveco Daily 49.10	Carlyle Dailybus 2	B25F	1989	
33	G533SBA	Iveco Daily 49.10	Carlyle Dailybus 2	DP23F	1989	
35	H835DNE	Iveco Daily 49.10	Phoenix	DP23F	1990	
36	H536CJA	Iveco Daily 49.10	Carlyle Dailybus 2	DP23F	1990	
37	H537CJA	Iveco Daily 49.10	Carlyle Dailybus 2	DP23F	1990	
38	PJI3738	Iveco Daily 49.10	Robin Hood City Nippy	B19F	1986	Ex GM Buses, 1992
42	F242FNE	Iveco Daily 49.10	Northern Counties	B22F	1988	
51	SCH151X	Leyland Leopard PSU3F/4R	Willowbrook 003	C49F	1981	Ex Trent, 1994
54	VNN54Y	Leyland Leopard PSU3F/4R	Willowbrook 003	C49F	1982	Ex Trent, 1994
71	ONN771M	Leyland Leopard PSU3B/4R	Plaxton Elite III Express	C53F	1974	Ex Barton, 1994
73	VXI8973	Leyland Leopard PSU5C/4R	Plaxton Supreme IV	C50F	1981	Ex Ribble, 1991
74	VXI8974	Leyland Leopard PSU5C/4R	Plaxton Supreme IV	C50F	1981	Ex Ribble, 1991
85	OIB1285	Leyland Tiger TRCTL11/2R	Plaxton Paramount 3200 E	C53F	1983	Ex Rossendale, 1993
86	J886PNC	Iveco Daily 49.10	Carlyle Dailybus 2	B23F	1991	
87	SJI1887	Freight Rover Sherpa	Carlyle Citybus 2	B21F	1988	Ex Daly Bus, Eccles, 1993
88	F288FLG	Iveco Daily 49.10	Northern Counties	B22F	1988	Ex Northern Counties demo, 1990
89	E189CNE	Iveco Daily 49.10	Northern Counties	B22F	1988	Ex GM Buses, 1989
98	F998XOV	Iveco Daily 49.10	Carlyle Dailybus 2	B23F	1988	Ex Strathclyde, 1990
99	F999XOV	Iveco Daily 49.10	Carlyle Dailybus 2	B23F	1988	Ex Strathclyde, 1990

Previous Registrations:

HSK330	KRN106T	PJI3738	D538MJA	VXI8974	MRJ274W
LBU607V	YHG16V	RIB1728	BNB235T	VXI8975	MRJ267W
LSU901	C453OFL	SJI1887	F225AKG	VXP717	A806CCD, 416CCD, A474NJK
OIB1285	A110EPA	VXI8973	MRJ273W		

Livery: Blue and white.

Bluebird have developed a network of services radiating from Middleton, a town north of Manchester, and now in the Rochdale District of Greater Manchester. Many of the population live on a large housing estate known as Langley, built in the 1950s to house people on the Manchester housing lists. Loading at Middleton bus station, above, bound for Langley is HSK330, while below is H537CJA, photographed working the Heywood circuar service 465. *Phillip Stephenson/Ken Grime*

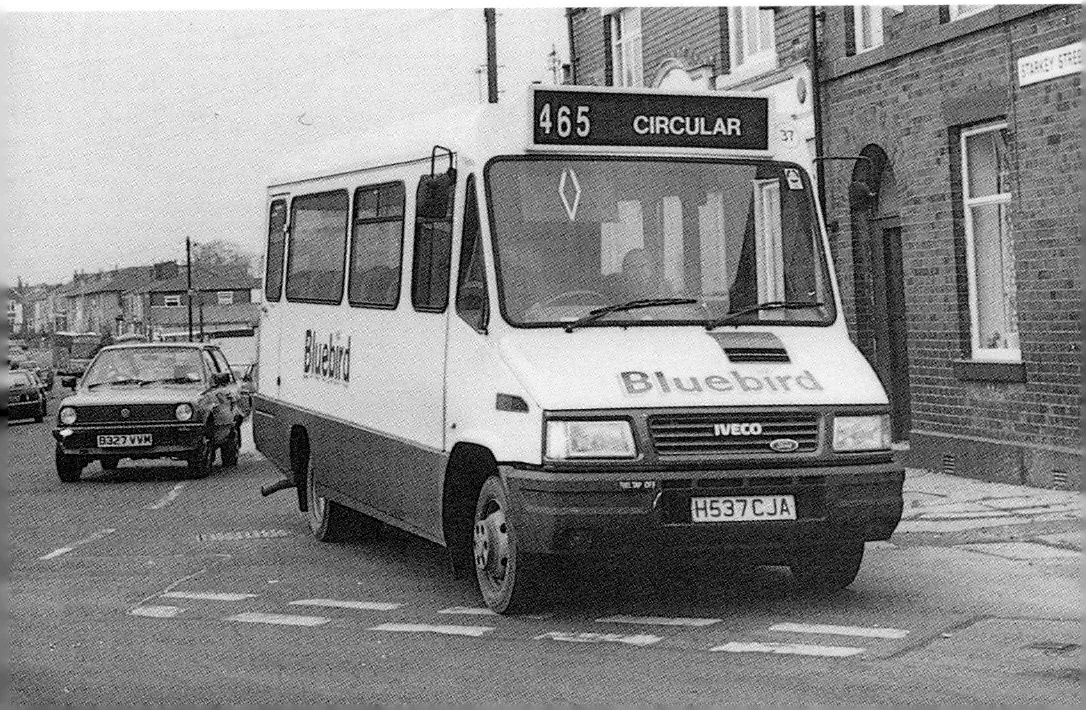

BLUE BUS

Blue Bus and Coach Services Ltd, Unit 4a, Locomotion Industrial Estate, Chorley New Road, Horwich, Bolton, Greater Manchester, BL6 5UE

Depots: Chorley New Road, Horwich and Canal Street, Huddersfield.

1	K1BLU	Dennis Dart 9.8SDL3003	East Lancashire	B40F	1993	
2	M2BLU	Dennis Dart 9.8SDL3003	East Lancashire	B40F	1993	
6w	NTX363R	Leyland Leopard PSU3C/2R	Willowbrook 01	B51F	1976	Ex Mayne, Manchester, 1992
7	OSJ607R	Leyland Leopard PSU3C/3R	Alexander AY	B53F	1976	Ex Leigh HGV Trainig, 1993
9	OEX799W	Leyland Leopard PSU3E/4R	Willowbrook 003	C49F	1980	Ex C & H, Fleetwood, 1991
10	DDM30X	Leyland Leopard PSU3E/4R	Willowbrook	C48F	1982	Ex Reynolds, Gwespyr, 1993
11	PRA115R	Leyland Leopard PSU3D/4R	Alexander AT	DP49F	1976	Ex South Lancs, St Helens, 1991
12	LHL246P	Leyland Leopard PSU3C/4R	Alexander AT	DP51F	1976	Ex Graham's, Stoke, 1993
14	PRA114R	Leyland Leopard PSU3D/4R	Alexander AT	DP49F	1976	Ex Vale, Cheetham, 1994
15	YSF99S	Leyland Leopard PSU3E/4R	Alexander AYS	B53F	1977	Ex Fife Scottish, 1993
16	ULS316T	Leyland Leopard PSU3E/4R	Alexander AYS	B53F	1979	Ex Rossendale, 1993
17	YSF97S	Leyland Leopard PSU3D/4R	Alexander AYS	B53F	1977	Ex Fife Scottish, 1992
18	ULS318T	Leyland Leopard PSU3E/4R	Alexander AYS	B53F	1979	Ex Rossendale, 1993
19	BWU691H	Leyland Leopard PSU4A/2R	Pennine	B43F	1969	Ex Shearings, 1991
20	SCS360M	Leyland Leopard PSU3/3R	Alexander AY	DP49F	1974	Ex Meredith, Malpas, 1993
22	ULS322T	Leyland Leopard PSU3E/4R	Alexander AYS	B53F	1979	Ex Rossendale, 1993
23	EAW623L	Leyland Leopard PSU3B/2R	Willowbrook	B53F	1972	Ex South Lancs, St Helens, 1991
24	ULS334T	Leyland Leopard PSU3E/4R	Alexander AYS	B53F	1979	Ex Rossendale, 1993
26	GMS276S	Leyland Leopard PSU3E/4R	Alexander AYS	B53F	1978	Ex Moffat & Williamson, Gauldry, 1994
27	GMS277S	Leyland Leopard PSU3E/4R	Alexander AYS	B53F	1978	Ex Moffat & Williamson, Gauldry, 1994
28	GMS294S	Leyland Leopard PSU3E/4R	Alexander AYS	B53F	1978	Ex Moffat & Williamson, Gauldry, 1994
29	GMS299S	Leyland Leopard PSU3E/4R	Alexander AYS	B53F	1978	Ex Moffat & Williamson, Gauldry, 1994
30w	WVM901S	Leyland Atlantean AN68A/1R	Park Royal	H43/32F	1978	Ex Pennine Blue, Denton, 1992
31	CWG691V	Leyland Atlantean AN68A/1RSp	Alexander AL	H45/29D	1979	Ex South Yorkshire, 1991
32	CWG683V	Leyland Atlantean AN68A/1RSp	Alexander AL	H45/29D	1979	Ex South Yorkshire, 1991
33	JFV313S	Leyland Atlantean AN68A/2R	East Lancashire	H50/36F	1978	Ex Blackpool, 1991
38	OIB9378	Leyland Atlantean AN68/1R	Alexander AL	H45/36F	1973	Ex Lancaster, 1991
39	OIB9379	Leyland Atlantean AN68/1R	Alexander AL	H45/36F	1973	Ex Lancaster, 1991
44	OWO234Y	Leyland Leopard PSU3G/2R	Duple Dominant	DP49F	1982	Ex Mayne, Manchester, 1993
51	LDZ2951	Leyland Leopard PSU3E/4R	East Lancs EL2000 (1992)	B51F	1980	Ex Trent, 1992
52	HIL9152	Leyland Leopard PSU3F/4R	East Lancs EL2000 (1992)	B51F	1982	Ex Trent, 1992
53	KWO263X	Leyland Leopard PSU3F/4R	East Lancs EL2000 (1994)	B51F	1982	Ex National Welsh, 1993
61	RJI2161	Leyland Tiger TRCTL11/2R	East Lancs EL2000 (1994)	B51F	1981	Ex Burman, Dorden 1994
71	PYD984V	Volvo B58-61	East Lancs EL2000 (1994)	B51F	1979	Ex Pride of the Road, 1994
85	KSU851P	Leyland Atlantean AN68A/1R	Alexander AL	H45/31F	1975	Ex ABC Travel, Ainsdale, 1992
89	KSA189P	Leyland Atlantean AN68A/1R	Alexander AL	H45/29D	1976	Ex G & M, Cefn Cribwr, 1992
100	K100BLU	DAF SB3000DKVF601	Van Hool Alizée	C55F	1993	
101	F701ENE	Leyland Tiger TRCL10/3ARZA	Plaxton Paramount 3200 III	C53F	1989	Ex Shearings, 1992
102	F702ENE	Leyland Tiger TRCL10/3ARZA	Plaxton Paramount 3200 III	C53F	1989	Ex Shearings, 1992

Previous Registrations:

EAW623L	KNY924L, 1810VT	LDZ2951	KVO143W	OIB9379	NRG162M
HIL9152	SCH150X	OIB9378	NRG158M	RJI2161	ESU157X, RDV903, HGD802X
K100BLU	K109TCP				

Livery: Blue and cream

Blue Bus of Horwich have recently introduced a service in Huddersfield, West Yorkshire, in order to expand services. One of the early vehicles transferred to Yorkshire is K1BLU, a Dennis Dart with East Lancashire bodywork. This vehicle is seen *(Above)* passing Bolton Odeon while operating service 573 from Bolton to Bury. *Below* is Leyland Atlantean 33, JFV313S, an example new to Blackpool. It was photographed at Ainsworth, approximately half way between Bolton and Bury on service 509 and once the terminus of the 9 in Bolton Transport days. *Roy Marshall/Ken Grime*

BLUE BUSES

Fylde Transport Ltd, Squires Gate, Lytham St Annes, Lancashire, FY8 2SH

1	H1FBT	DAF SB220LC550	Optare Delta	DP48F	1991	
2	H2FBT	DAF SB220LC550	Optare Delta	DP48F	1991	
3	H3FBT	DAF SB220LC550	Optare Delta	DP48F	1991	
4	TKU462K	Leyland Atlantean AT68M/2RF	N Counties Paladin (1992)	DP42F	1971	Ex Kingston-upon-Hull, 1987
5	TKU465K	Leyland Atlantean AT68M/2RFT	N Counties Paladin (1992)	DP42F	1971	Ex Kingston-upon-Hull, 1987
6	TKU466K	Leyland Atlantean AT68M/2RFT	N Counties Paladin (1992)	DP42F	1971	Ex Kingston-upon-Hull, 1987
7	TKU469K	Leyland Atlantean AT68M/2RFT	N Counties Paladin (1992)	DP42F	1971	Ex Kingston-upon-Hull, 1987
13	B43UCK	Leyland Tiger TRBTL11/2RH	Duple Dominant	DP47F	1984	
24	G812RNC	Leyland Tiger TR2R62C21Z5	Plaxton Paramount 3200 III	C53F	1990	Ex Shearings, 1993
25	G813RNC	Leyland Tiger TR2R62C21Z5	Plaxton Paramount 3200 III	C53F	1990	Ex Shearings, 1993
26	J26LRN	DAF SB3000DKV601	Van Hool Alizée	C51FT	1992	
27	MJI7846	Leyland Tiger TRCTL11/3RH	Duple Laser 2	C53F	1985	
28	TRN772	Leyland Leopard PSU3C/4R	Duple 320 (1989)	C53F	1977	
29	YNJ434	Leyland Leopard PSU3C/4R	Duple 320 (1990)	C53F	1975	
30	NUT16W	Leyland Leopard PSU3F/5R	Duple Dominant II	C53F	1981	Ex Wood, Blackpool, 1988
31	SIB4631	Volvo B58-56	Duple Dominant II	C53F	1980	
32	PJI5632	Leyland Leopard PSU3B/4R	Duple Dominant	C53F	1974	Ex Wood, Blackpool, 1988
33	OHG33T	Leyland Leopard PSU3E/4R	Duple Dominant I	C53F	1979	
34	OHG34T	Leyland Leopard PSU3E/4R	Duple Dominant I	C53F	1979	
36	XRN29V	Leyland Leopard PSU3E/4R	Plaxton Supreme IV Express	C53F	1980	
41	HIL5341	Leyland Atlantean AN68C/2R	Northern Counties	DPH43/33F	1981	
42	HIL5342	Leyland Atlantean AN68C/2R	Northern Counties	DPH43/33F	1982	
43	HIL5943	Leyland Atlantean AN68D/2R	Northern Counties (1992)	DPH43/33F	1983	
44	NJI5504	Leyland Atlantean AN68D/2R	Northern Counties	DPH43/33F	1983	
45	NJI5505	Leyland Atlantean AN68D/2R	Northern Counties	DPH43/33F	1984	

Opposite, top: **In additon to double-deck rebodies, four Atlantean chassis were totally rebuilt and new Northern Counties Paladin bodies added. The last vehicle to be completed was 7, TKU469K, seen working service 11.** *Paul Wigan*

Opposite, bottom: **Blue Bus 58, ATD281J, is seen on Blackpool Promanade in its open-top configuration.** *Paul Wigan*

In 1990, Fylde started a programme to rebulid Leyland Atlanteans. This involved not just new Northern Counties bodies. For three examples this resulted in a change of chassis designation, as shown in the fleetlist. The first of the rebuilds is 71, OJI4371. *Roy Marshall*

11
LYTHAM SQUARE
via St. Annes
BLUE BUSES
the FYLDE's own TRANSPORT
7
BLUE BUSES
the FYLDE's own TRANSPORT
TKU 469K

EVERY FEW MINUTES
EVERY DAY
ON THIS BUS
ROLLER COASTER
ROUTES 1 & 1A
PROMENADE BUS
STARR GATE SANDCASTLE
3 PIERS TOWER GYNN SQ
WELCOME ABOARD
1 ST. ANNES
PEPSI MAX BIG ONE • VERTICAL REALITY ARRIVES IN BLACKPOOL...
THE WORLD'S TALLEST & FASTEST ROLLER COASTER!
58
BLUE BUSES
ATD 281J
TOWER

48	ARH304K	Leyland Atlantean PDR1A/1	Roe	H43/29F	1971	Ex Kingston-upon-Hull, 1986
49	ARH314K	Leyland Atlantean PDR1A/1	Roe	H43/29F	1972	Ex Kingston-upon-Hull, 1988
50	HRN100N	Leyland Atlantean AN68/1R	NCounties/Willowbrook	H43/31F	1975	
51	DBA227C	Leyland Atlantean PDR1/1	MCW	O43/33F	1965	Ex Lancaster, 1986
52	TKH266H	Leyland Atlantean PDR1A/1	Roe	O43/28F	1969	Ex Kingston-upon-Hull, 1987
53	WRH294J	Leyland Atlantean PDR1A/1	Roe	O43/28F	1970	Ex Kingston-upon-Hull, 1987
54	ARH309K	Leyland Atlantean PDR1A/1	Roe	O43/29F	1971	Ex Kingston-upon-Hull, 1987
55	TKH268H	Leyland Atlantean PDR1A/1	Roe	O43/28F	1969	Ex Kingston-upon-Hull, 1987
56	WRH291J	Leyland Atlantean PDR1A/1	Roe	H43/28F	1970	Ex Kingston-upon-Hull, 1987
57	WRH295J	Leyland Atlantean PDR1A/1	Roe	H43/28F	1970	Ex Kingston-upon-Hull, 1987
58	ATD281J	Leyland Atlantean PDR1A/1	Northern Counties	H44/33F	1970	
59	ARH301K	Leyland Atlantean PDR1A/1	Roe	H43/29F	1971	Ex Kingston-upon-Hull, 1988
60	ARH308K	Leyland Atlantean PDR1A/1	Roe	H43/29F	1971	Ex Kingston-upon-Hull, 1988

61-69 Leyland Atlantean AN68A/1R — Northern Counties — H43/32F — 1976-77 — Ex GM Buses, 1989

61	ONF659R	**63**	ONF664R	**65**	ONF667R	**67**	ONF673R	**69**	SRJ757R
62	ONF660R	**64**	ONF666R	**66**	ONF669R	**68**	SRJ756R		

70	TSD571S	Leyland Atlantean AN68A/1R	Northern Counties	H44/34F	1978	Ex Clyde Coast, Ardrossan, 1990
71	OJI4371	Leyland Atlantean AN68A/1R	Northern Counties (1990)	H43/31F	1977	
72	OJI4372	Leyland Atlantean AN68M/1RF	Northern Counties (1991)	H43/31F	1977	
73	OJI4373	Leyland Atlantean AN68M/1RF	Northern Counties (1991)	H43/31F	1977	
74	OJI4374	Leyland Atlantean AN68A/1R	Northern Counties (1992)	H43/31F	1977	
75	SIB8405	Leyland Atlantean AN68M/1RFT	Northern Counties (1993)	H43/31F	1976	
78	HRN98N	Leyland Atlantean AN68/1R	NCounties/Willowbrook	H43/31F	1975	
79	HRN99N	Leyland Atlantean AN68/1R	NCounties/Willowbrook	H43/31F	1975	
81	HRN101N	Leyland Atlantean AN68/1R	NCounties/Willowbrook	H43/31F	1975	
82	HRN102N	Leyland Atlantean AN68/1R	NCounties/Willowbrook	H43/31F	1975	
83	HRN103N	Leyland Atlantean AN68/1R	NCounties/Willowbrook	H43/31F	1975	
86	MIW8186	Leyland Atlantean PDR1A/1	Roe	H43/29F	1971	Ex Kingston-upon-Hull, 1987
87	MIW8187	Leyland Atlantean AN68/1R	Roe	H43/29F	1973	Ex Kingston-upon-Hull, 1992
88	MIW8188	Leyland Atlantean AN68/1R	Roe	H43/29F	1972	Ex Kingston-upon-Hull, 1988
89	RIB4089	Leyland Atlantean AN68/1R	Roe	H43/29F	1972	Ex Kingston-upon-Hull, 1988
90	RIB6590	Leyland Atlantean PDR1A/1	Roe	H43/29F	1971	Ex Kingston-upon-Hull, 1987
91	IIL4291	Leyland Atlantean PDR1A/1	Roe	H43/29F	1971	Ex Kingston-upon-Hull, 1992
92	NIW6492	Leyland Atlantean AN68/1R	Roe	H43/29F	1972	Ex Kingston-upon-Hull, 1992
93	ARH306K	Leyland Atlantean PDR1A/1	Roe	H43/29F	1971	Ex Kingston-upon-Hull, 1992
95	RHG95T	Leyland Atlantean AN68A/1R	Northern Counties	H43/31F	1979	
96	XHG96V	Leyland Atlantean AN68A/1R	Northern Counties	H43/31F	1980	

101-118 Renault-Dodge S56 — Northern Counties — B25F — 1987-88 — 107 Ex Northern Counties, 1987
*105/6 are DP20F, 101-4/7 are B22F, 117/8 are B23F

101	D101AFV	**104**	D104AFV	**107**	D912NBA	**111**	E111LCW	**114**	E114LCW
102	D102AFV	**105**	D105AFV	**108**	E108LCW	**112**	E112LCW	**117**	E164CNC
103	D103AFV	**107**	D106AFV	**110**	E110LCW	**113**	E113LCW	**118**	E165CNC

120-125 Renault-Dodge S56 — Northern Counties — B27F — 1988 — *120 is DP27F

120	F120UFR	**122**	F142UFR	**123**	F143UFR	**124**	F144UFR	**125**	F145UFR
121	F141UFR								

134	BTB928	Leyland Lion LT7C	Leyland	B34R	1936	Re-acquired 1988

Previous Registrations:

HIL5341	HBV97W	NIW6492	DRH327L	RIB4089	DRH329L
HIL5342	PCW98X	NJI5504	A74LHG	RIB6590	ARH311K
HIL5943	ACK99Y	NJI5505	B75URN	SIB4631	XRN28V
IIL4291	ARH298K	OJI4371	EBV85S	SIB8405	OCK84P
MIW8186	ARH309K	OJI4372	EBV86S	TRN772	XRN36R
MIW8187	NAT339M	OJI4373	EBV87S	YNJ434	HRN97N
MIW8188	DRH330L	OJI4374	EBV88S		
MJI7846	C27ECW	PJI5632	YNA400M		

Livery: Fresco and Fylde blue and white

BORDER

P Cartmell, Heasandford Villa, Burnley, Lancashire, BB10 2AH

31	KSU857P	Leyland Atlantean AN68/1R	Alexander AL	H45/31F	1976	Ex Rossendale, 1993
39	PRJ485R	Daimler Fleetline CRG6LXB	Northern Counties	H43/32F	1976	Ex Rossendale, 1993
40	KSJ940P	Leyland Leopard PSU3C/4R	Alexander AYS	B53F	1976	Ex Western Scottish, 1993
44	SRJ744R	Leyland Atlantean AN68A/1R	Northern Counties	H43/32F	1977	Ex Yorkshire Rider, 1993
64	Q364FVT	Leyland Leopard PSU3A/4RT	Willowbrook Warrior (1992)	B48F	1970	Ex Lucas, Kingsley, 1989
74	RRC485R	Leyland Leopard PSU3C/4R	Plaxton Supreme III Express	C53F	1976	Ex Hyndburn, 1993
82	LIB1182	Leyland Leopard PSU3F/4R	Plaxton Supreme IV Express	C53F	1981	Ex Rossendale, 1989
113	CAU113T	Leyland Fleetline FE30ALR	Northern Counties	H44/31F	1979	Ex Nottingham, 1994
161	NRG161M	Leyland Atlantean AN68/1R	Alexander AL	H45/36F	1974	Ex Blue Bus, Horwich, 1993
165	NRG165M	Leyland Atlantean AN68/1R	Alexander AL	H45/36F	1974	Ex Blue Bus, Horwich, 1993
167	NRG167M	Leyland Atlantean AN68/1R	Alexander AL	H45/36F	1974	Ex Blue Bus, Horwich, 1993
178	UDT178S	Leyland Atlantean AN68A/1R	East Lancashire	H45/31F	1978	Ex Mancunian, Manchester, 1994
311	LCB652P	Leyland Leopard PSU5A/4RT	Plaxton Supreme III	C53F	1976	Ex Rossendale, 1993
413	PDK829M	Leyland Leopard PSU5/4RT	Duple Dominant	C57F	1974	Ex Rossendale, 1993
1788	OEM788S	Leyland Atlantean AN68/1R	MCW	H43/32F	1978	Ex Merseybus, 1993
1811	PKA725S	Leyland Atlantean AN68/1R	MCW	H43/32F	1978	Ex Merseybus, 1993

Previous Registrations:

Q364FVT	LJX817H	LIB1182	GBO245W

Livery: Red and white

Two leyland Atlanteans with MCW bodywork joined the Border fleet in 1993, previously having been with the Mersybus fleet. It is seen while working service B7. *Ken Grime*

BULLOCK

R Bullock & Co (Transport) Ltd, Commercial Garage, Stockport Road, Cheadle, Greater Manchester, SK8 2AG.

WBN998L	Daimler Fleetline CRG6LXB	Park Royal	H43/32F	1973	Ex Greater Manchester, 1988
YNA293M	Daimler Fleetline CRG6LXB	Northern Counties	H43/32F	1973	Ex Greater Manchester, 1987
YNA294M	Daimler Fleetline CRG6LXB	Northern Counties	H43/32F	1973	Ex Greater Manchester, 1987
YNA354M	Daimler Fleetline CRG6LXB	Northern Counties	H43/32F	1973	Ex Finglands, Manchester, 1994
BNE736N	Daimler Fleetline CRG6LXB	Northern Counties	H43/32F	1974	Ex Greater Manchester, 1987
HFM804N	Leyland Leopard PSU5/4R	Plaxton Elite III	C57F	1978	
KUC216P	Daimler Fleetline CRL6	Park Royal	H44/27D	1976	Ex Grimsby-Cleethorpes, 1991
OJD249R	Leyland Fleetline FE30ALR Sp	MCW	H44/24D	1979	Ex Network Express, Maidenhead, 1993
RCH282R	Leyland Fleetline FE30AGR	Roe	H43/30F	1976	Ex Derby, 1990
OHR182R	Leyland Fleetline FE30AGR	Eastern Coach Works	H43/31F	1977	Ex Wealdon, Oak Green, 1991
SUA9R	Leyland Fleetline FE30AGR	Northern Counties	H43/31F	1977	Ex Yorkshire Rider, 1990
AFA489S	Leyland Fleetline FE30AGR	Northern Counties	H44/33F	1978	Ex PMT, 1989
THX272S	Leyland Fleetline FE30ALR Sp	MCW	H44/27D	1978	Ex London Buses, 1991
THX539S	Leyland Fleetline FE30ALR Sp	Park Royal	H44/29D	1978	Ex London Buses, 1991
THX563S	Leyland Fleetline FE30ALR Sp	Park Royal	H44/27F	1978	Ex London Buses, 1991
WUM107S	Leyland Fleetline FE30AGR	Roe	H43/33F	1978	Ex Yorkshire Rider, 1990
AYR347T	Leyland National 10351A/2R		B36D	1978	Ex London Buses, 1990
ULS665T	Leyland Fleetline FE30AGR	Eastern Coach Works	H43/32F	1979	Ex Clydeside 2000, 1993
LVT699V	Leyland Fleetline FE30AGR	Northern Counties	H44/33F	1980	Ex PMT, 1989
BYW362V	Leyland National 10351A/2R		B44F	1980	Ex Rose Hill, Marple, 1994
BUI1675	Leyland Leopard PSU3F/4R	East Lancashire (1993)	B51F	1981	Ex Go-Ahead Northern, 1990
WDB390X	Leyland Leopard PSU5D/5R	Plaxton Supreme V	C57F	1982	
CKS385X	MCW Metrobus DR102/24	Alexander RL	H45/33F	1982	Ex BTS, Borehamwood, 1992
ULS616X	MCW Metrobus DR102/28	Alexander RL	H45/33F	1982	Ex BTS, Borehamwood, 1992
ULS617X	MCW Metrobus DR102/28	Alexander RL	H45/33F	1982	Ex BTS, Borehamwood, 1992
BUI1424	Dennis Dorchester SDA805	Berkhof Esprite 350	C53F	1983	Ex Berkhof demonstrator, 1984
BUI1610	Dennis Dorchester SDA805	Plaxton Paramount 3200	C57F	1984	
A301JFA	Leyland National 2 NL116HLXCT/1R		DP47F	1984	Ex Lawrenson, Haydock, 1994
A303JFA	Leyland National 2 NL116HLXCT/1R		DP47F	1984	Ex Lawrenson, Haydock, 1994
BUI1133	Dennis Dorchester SDA805	Plaxton Paramount 3200	C57F	1985	
BUI1484	Leyland Tiger TRCLXCT/3RZ	Plaxton P'mount 3500(1983)	C51FT	1986	
C281BBP	Dennis Dominator DDA1008	East Lancashire	H45/31F	1986	Ex Southampton Citybus, 1991
C282BBP	Dennis Dominator DDA1008	East Lancashire	H45/31F	1986	Ex Southampton Citybus, 1991
C283BBP	Dennis Dominator DDA1008	East Lancashire	H45/31F	1986	Ex Southampton Citybus, 1991
C284BBP	Dennis Dominator DDA1008	East Lancashire	H45/31F	1986	Ex Southampton Citybus, 1991
C285BBP	Leyland Olympian ONLXB/1R	East Lancashire	DPH43/27F	1986	Ex Southampton Citybus, 1992

Bullock have a major presence on services to the Didsbury area of south Manchester. C284BBP, a Dennis Dominator with East Lancashire bodywork, crosses the tram lines in Piccadilly heading for Parswood, East Didsbury.
Keith Grimes

The Bullock fleet contains seven former London DMSs, here represented by THX272S, photographed in Piccadilly, Manchester. Service 42 is one of many operating along Wilmslow Road.

D391BNR	Ford R1014	Plaxton Paramount 3200 II	C35FT	1987	
E419YLG	Mercedes-Benz 609D	Made-to-Measure (1991)	C24F	1987	Ex van, 1991
E758OWY	Mercedes-Benz 709D	Made-to-Measure (1993)	B20F	1988	Ex van, 1993
E393SWX	Mercedes-Benz 811D	Olympus (1994)	B21F	1988	Ex van, 1994
E161TVR	Leyland Tiger TRCTL11/3ARZ	Duple 340	C55F	1988	
F450FDB	Leyland Tiger TRCTL11/3ARZ	Duple 320	C57F	1989	
F452FDB	Leyland Tiger TRBTL11/2RP	Duple 300	DP55F	1989	
G481HNP	Mercedes-Benz 811D	Olympus (1994)	B23F	1990	Ex van, 1994
BUI1300	DAF SB3000DKV601	Caetano Algarve	C50FT	1990	Ex Boyden, Castle Donington, 1991
G422SNF	Dennis Javelin 12SDA1907	Duple 320	C57F	1990	
G423SNF	Leyland Tiger TRBTL11/2RP	Duple 300	DP55F	1990	
H10BUL	Volvo B10M-60	Plaxton Paramount 3500 III	C53F	1991	Ex Park's, Hamilton, 1992
J100BUL	Dennis Javelin 12SDA1919	Duple 320	C57F	1991	
J200BUL	Leyland Tiger TRCL10/3ARZA	Plaxton Paramount 3200 III	C57F	1991	
J400BUL	Volvo B10M-60	Plaxton Paramount 3500 III	C53F	1992	
K500BUL	Scania K93CRB	Plaxton Paramount 3500 III	C55F	1993	
K600BUL	Volvo B10M-60	Plaxton Paramount 3200 III	C53F	1993	
K700BUL	Dennis Javelin 11SDA1906	Duple 320	C53F	1993	
L10BUL	Leyland Olympian ON2R50C13Z4	East Lancashire	H45/31F	1993	
L20BUL	Leyland Olympian ON2R50C13Z4	East Lancashire	H45/31F	1993	
L800BUL	Volvo B10M-60	Van Hool Alizée	C51FT	1994	

Previous Registrations:

BUI1133	B593SNC	BUI1484	C33EVM	J400BUL	J729KBC
BUI1300	G159XJF	BUI1610	A530OBU	K600BUL	K433ANE
BUI1424	KNO224Y	BUI1675	FNL687W	K700BUL	K326ANC

Livery: Red (buses); white, red and gold (coaches)

BURNLEY & PENDLE

Burnley & Pendle Transport Co, Queensgate, Colne Road, Burnley,
Lancashire, BB10 1HH

2	GSU552	Volvo B10M-61	Jonckheere Jubilee P50	C49F	1986	
3	GSU553	Volvo B10M-61	Jonckheere Jubilee P50	C49F	1986	
4	GSU554	Leyland Tiger TRCTL11/3R	Duple Laser 1	C53F	1983	Ex GM Buses, 1989
5	XSU905	Leyland Tiger TRCTL11/3R	Duple Laser 1	C55F	1983	Ex GM Buses, 1989
6	XSU906	Mercedes-Benz 709D	Reeve Burgess Beaver	C19F	1987	
7	XSU907	Volvo C10M-70	Ramseier & Jenzer	C49FT	1985	Ex Park's, Hamilton, 1990
8	XSU908	Volvo C10M-70	Ramseier & Jenzer	C49FT	1985	Ex Park's, Hamilton, 1990
9	XSU909	Volvo B10M-61	Jonckheere Jubilee P50	C51FT	1986	Ex Parry, Cheslyn Hay, 1990
10	XSU910	Volvo B10M-61	Jonckheere Jubilee P50	C49FT	1987	Ex Parry, Cheslyn Hay, 1990
11	HXI311	Volvo B10M-61	Jonckheere Jubilee P50	C49FT	1988	Ex Parry, Cheslyn Hay, 1991
12	NXI812	Volvo B10MT-53	Van Hool Alizée	C48FT	1989	Ex Rothwell, Heywood, 1991
13	GXI613	Volvo B10MT-50	Plaxton Paramount 4000	CH53/12DT	1989	Ex Flight's, Birmingham, 1992
14	NXI414	Volvo B10MT-53	Jonckheere Jubilee P95	CH54/13DT	1985	Ex Flight's, Birmingham, 1988
15	K815TJU	Toyota HDB30R	Caetano Optimo	C21F	1992	

17-23 Volvo Citybus B10M-55 — Alexander PS — B51F — 1991

17	H617ACK	**19**	H619ACK	**21**	H621ACK	**22**	H622ACK	**23**	H623ACK
18	H618ACK	**20**	H620ACK						

24	J24MCW	Volvo Citybus B10M-50	East Lancashire EL2000	B45F	1992
25	J25MCW	Volvo Citybus B10M-50	East Lancashire EL2000	B45F	1992
26	K26WBV	Volvo Citybus B10M-50	East Lancashire EL2000	B45F	1993
27	K27WBV	Volvo Citybus B10M-50	East Lancashire EL2000	B45F	1993

Tri-axle chassis are needed to meet certain axle weight regulations - especially for double-deck vehicles. Three are in the Burnley & Pendle fleet, including 14, NXI414, seen leaving Burnley bus station. This example is fitted with a Jonckheere Jubilee P95 body. *Roy Marshall*

One of the latest single-deck buses for Burnley & Pendle is 27, K27WBV, a Volvo Citybus B10M with East Lancashire bodywork. *Ken Grime*

30-43		Leyland Leopard PSU4E/4R		East Lancashire		B47F	1978-80		
30	MFV30T	**33**	MFV33T	**36**	MFV36T	**39**	DBV39W	**42**	DBV42W
31	MFV31T	**34**	MFV34T	**37**	MFV37T	**40**	DBV40W	**43**	DBV43W
32	MFV32T	**35**	MFV35T	**38**	DBV38W	**41**	DBV41W		

44-50		Leyland National 2 NL116L11/1R				B52F	1980		
44	XRN44V	**46**	XRN46V	**48**	XRN48V	**49**	XRN49V	**50**	XRN50V
45	XRN45V	**47**	XRN47V						

51	GSU551	Leyland Tiger TRCTL11/3R	Plaxton Paramount 3500	C53F	1983	
57	VFV7V	Leyland Leopard PSU3E/4R	Duple Dominant II	C53F	1979	
58	VFV8V	Leyland Leopard PSU3E/4R	Duple Dominant II	C53F	1979	

61-66		Volvo Citybus B10M-55		Alexander P		DP53F	1988		
61	E61JFV	**63**	E63JFV	**64**	E64JFV	**65**	E65JFV	**66**	E66JFV
62	E62JFV								

67	G67PFR	Volvo Citybus B10M-55	East Lancashire EL2000	B51F	1990	
68	G68PFR	Volvo Citybus B10M-55	East Lancashire EL2000	B51F	1990	
69	BUH239V	Leyland National 2 NL106L11/1R		B44F	1980	Ex National Welsh, 1988
70	BUH240V	Leyland National 2 NL106L11/1R		B44F	1980	Ex National Welsh, 1989
71	BUH241V	Leyland National 2 NL106L11/1R		B44F	1980	Ex National Welsh, 1988
72	FUH32V	Leyland National 2 NL116L11/1R		B49F	1980	Ex Taff Ely, 1988
73	FUH33V	Leyland National 2 NL116L11/1R		B49F	1980	Ex Taff Ely, 1988
74	FUH34V	Leyland National 2 NL116L11/1R		B49F	1980	Ex Taff Ely, 1988
75	K75XCW	Optare MetroRider MR03	Optare	B29F	1993	
76	J176MCW	Optare MetroRider MR09	Optare	B23F	1992	
77	J177MCW	Optare MetroRider MR09	Optare	B23F	1992	
78	H78CFV	Mercedes-Benz 811D	Alexander AM	B31F	1991	
79	H79CFV	Mercedes-Benz 811D	Alexander AM	B31F	1991	
80	C80OCW	Mercedes-Benz L608D	Reeve Burgess	B20F	1986	
81	D81UFV	Mercedes-Benz L608D	Sparshatts	B20F	1986	
82	D82UFV	Mercedes-Benz L608D	Sparshatts	B20F	1986	
83	D83UFV	Mercedes-Benz L608D	Sparshatts	B20F	1986	
84	E84HRN	Mercedes-Benz 709D	Robin Hood	B29F	1987	

85-90		Mercedes-Benz 811D		Robin Hood		B29F	1987-88		
85	E85HRN	**87**	E87HRN	**88**	E88HRN	**89**	E89HRN	**90**	E90JHG
86	E86HRN								

91	E91LBV	Mercedes-Benz 709D	Alexander AM	B25F	1988	
92	E92LHG	Mercedes-Benz 709D	Alexander AM	B25F	1988	
93	E93LHG	Mercedes-Benz 709D	Alexander AM	B25F	1988	
94	E94LHG	Mercedes-Benz 709D	Alexander AM	B25F	1988	
95	XFK305	Mercedes-Benz 709D	Reeve Burgess Beaver	DP25F	1988	
96	G96MRN	Mercedes-Benz 811D	Reeve Burgess Beaver	B31F	1990	
97	G97MRN	Mercedes-Benz 811D	Reeve Burgess Beaver	B31F	1990	
98	G98PCK	Mercedes-Benz 811D	Reeve Burgess Beaver	B31F	1990	
99	G99PCK	Mercedes-Benz 709D	Reeve Burgess Beaver	B23F	1990	
101	E101JFV	Volvo Citybus B10M-50	Alexander RV	DPH47/35F	1988	
102	E102JFV	Volvo Citybus B10M-50	Alexander RV	DPH47/35F	1988	

103-112		Volvo Citybus B10M-50		Alexander RV		H47/37F	1989		
103	F103XCW	**105**	F105XCW	**107**	F107XCW	**109**	F109XCW	**111**	F111XCW
104	F104XCW	**106**	F106XCW	**108**	F108XCW	**110**	F110XCW	**112**	F112XCW

113	H113ABV	Volvo Citybus B10M-50	Alexander RV	H47/37F	1991
114	H114ABV	Volvo Citybus B10M-50	Alexander RV	DPH47/35F	1991
115	H115ABV	Volvo Citybus B10M-50	Alexander RV	DPH47/35F	1991

166-174		Bristol VRT/SL3/6LXB		Eastern Coach Works		H43/31F	1978		
166	FFR166S	**168**	FFR168S	**172**	FFR172S	**173**	FFR173S	**174**	FFR174S
167	FFR167S								

178	L178KHG	Optare MetroRider MR17	Optare	B29F	1994
179	L179KHG	Optare MetroRider MR17	Optare	B29F	1994

193-205		Bristol VRT/LL3/6LXB		Alexander AL		H49/35F	1977	Ex Tayside, 1982	
193	OSR193R	**195**	OSR195R	**196**	OSR196R	**197**	OSR197R	**205**	OSR205R

325	HHG25	Leyland Tiger Cub PSUC1/1	East Lancashire	B41F	1959

Previous Registrations:

GSU551	A201MFR	HXI311	E508KNV	XSU906	E206FRN
GSU552	D202VBV	NXI414	B708EOF, HHG25	XSU907	C641KDS
GSU553	D203VBV	NXI812	G827UMU	XSU908	C661KDS
GSU554	ANA52Y	XFK305	F95VBV	XSU909	D111BNV
GXI613	F705COA	XSU905	ANA53Y, XFK305	XSU910	E507KNV
HHG25	HHG25, MBV402G				

Livery: Crimson and cream; white grey and maroon (Viscount Central) 2-15, 51

Representing Burnley & Pendle's double decks is 101, E101JFV, a Volvo B10M with Alexander RV-type bodywork. This vehicle's high-back seating was most useful for a journey to Wembley.
Keith Grimes

BU-VAL

M Bull, Unit 5, Paragon Industrial Estate, Smithybridge, Littleborough,
Greater Manchester, OL15 8QF

D681SEM	Renault-Dodge S56	Alexander AM	B23F	1987	Ex Merseybus, 1994
D692SEM	Renault-Dodge S56	Alexander AM	B23F	1987	Ex Merseybus, 1994
D699THF	Renault-Dodge S56	Alexander AM	B23F	1987	Ex Merseybus, 1994
D979TKC	Renault-Dodge S56	Alexander AM	B23F	1987	Ex Merseybus, 1994
D905MDB	Renault-Dodge S56	Northern Counties	B20F	1987	Ex Merseybus, 1994
E702UEM	Renault-Dodge S56	Alexander AM	B23F	1987	Ex Merseybus, 1994
E705UEM	Renault-Dodge S56	Alexander AM	B23F	1987	Ex Merseybus, 1994
E181CNE	Iveco Daily 49.10	Northern Counties	B22F	1988	Ex GM Buses, 1989
E186CNE	Iveco Daily 49.10	Northern Counties	B22F	1988	Ex GM Buses, 1989
E187CNE	Iveco Daily 49.10	Northern Counties	B22F	1988	Ex GM Buses, 1989
F226FNE	Iveco Daily 49.10	Northern Counties	B24F	1988	
F241FNE	Iveco Daily 49.10	Northern Counties	B22F	1988	
F243FNE	Iveco Daily 49.10	Northern Counties	B22F	1988	
L803FBA	Iveco TurboDaily 59.12	Mellor	B27F	1994	
L804FBA	Iveco TurboDaily 59.12	Mellor	B27F	1994	

Livery: Red, white and grey.

The latest arrival with Bu-Val is L804FBA, an Iveco TurboDaily 59.12 with Mellor bodywork. Mellor are based in nearby Rochdale where, in addition to PCV minibuses the firm build welfare bodies in large numbers.

C M T

C & M Travel, Unit D3, Liver Industrial Estate, Long Lane, Aintree, Liverpool, Merseyside, L4 7ES

1002	PTT79R	Leyland National 11351A/1R		B53F	1976	Ex Wingates Tours, Melling, 1993
1013	B868UFR	Mercedes-Benz L307D	Imperial	M12	1984	
1020	JBO346N	Leyland National 10351/2R		B44F	1974	Ex Cardiff, 1989
1021	JBO347N	Leyland National 10351/2R		B44F	1974	Ex Cardiff, 1989
1022	JBO350N	Leyland National 10351/2R		B40D	1974	Ex Cardiff, 1990
1023	JBO351N	Leyland National 10351/2R		B40D	1974	Ex Cardiff, 1991
1025	BCD803L	Leyland National 1151/1R/0102		B49F	1973	Ex Southdown, 1989
1028	AYR321T	Leyland National 10351A/2R		B36D	1977	Ex Pennine Blue, 1992
1029	AYR346T	Leyland National 10351A/2R		B36D	1977	Ex Pennine Blue, 1992
1031	CNB253M	Leyland National 11351/1R/SC		DP48F	1974	Ex Wingates Tours, Melling, 1990
1033	BVU917N	Leyland National 10351/1R		B41F	1974	Ex Wingates Tours, Melling, 1990
1034	KNH500N	Leyland National 11351/1R		B49F	1975	Ex United Counties, 1990
1035	THX253S	Leyland National 10351A/2R		B36D	1976	Ex Pennine Blue, 1991
1036	HSC110T	Leyland National 11351A/1R		B49F	1978	Ex Wigan Bus Company, 1994
1037	HSC113T	Leyland National 11351A/1R		B49F	1978	Ex Wigan Bus Company, 1994
1038	KJD516P	Leyland National 10351A/2R		B20DL	1975	Ex Armstrong, Bletchley, 1989
1061	WPT712R	Leyland National 11351A/1R		B49F	1977	Ex ?, 1994
1062	UFG61S	Leyland National 11351A/2R		B44D	1976	Ex Hants & Dorset, 1992
1063	WYJ170S	Leyland National 11351A/2R		B44D	1976	Ex Southdown, 1992
1064	YCD72T	Leyland National 11351A/1R		B49F	1976	Ex Southdown, 1992
1065	GTL359N	Leyland National 11351/1R		B49F	1975	Ex Pennine Blue, 1992
1066	RUF45R	Leyland National 11351A/2R		B44D	1976	Ex Hastings & District, 1992
1067	ABA29T	Leyland National 11351A/1R		B49F	1977	Ex Blue Triangle, Bootle, 1993
1068	YPL413T	Leyland National 10351B/1R		B41F	1978	Ex Blue Triangle, Bootle, 1993
1069	NPK232R	Leyland National 10351A/1R		B41F	1976	Ex County Bus, 1993
1070	LPB221P	Leyland National 10351/1R		B41F	1976	Ex County Bus, 1993
1071	NPK249R	Leyland National 10351A/1R		B41F	1976	Ex County Bus, 1993

C & M Travel have a fleet almost entirely comprising Leyland Nationals. Seen passing St John's shopping centre is 1066, RUF45R. *Roy Marshall*

CMT's 1037, HSC113T was new to Alexander (Fife) in 1974, but was bought from the Wigan Bus Company some 20 years later. It is seen here working service 345 to Lydgate. *Phillip Stephenson*

1072	SPC266R	Leyland National 10351A/1R	B41F	1977	Ex County Bus, 1993
1073	SPC284R	Leyland National 10351A/1R	B41F	1977	Ex County Bus, 1993
1074	SPC286R	Leyland National 10351A/1R	B41F	1977	Ex County Bus, 1993
1075	SPC291R	Leyland National 10351A/1R	B41F	1977	Ex County Bus, 1993
1076	SPC278R	Leyland National 10351A/1R	B41F	1977	Ex County Bus, 1993
1077	SPC267R	Leyland National 10351A/1R	B41F	1977	Ex County Bus, 1993
1078	LPB222P	Leyland National 10351/1R	B41F	1976	Ex County Bus, 1993
1079	UPB309S	Leyland National 10351A/1R	B41F	1977	Ex County Bus, 1993
1080	UPB317S	Leyland National 10351A/1R	B41F	1977	Ex County Bus, 1993
1081	LPB212P	Leyland National 10351/1R	B41F	1976	Ex County Bus, 1993
1082	UPB308S	Leyland National 10351A/1R	B41F	1977	Ex County Bus, 1993
1083	THX166S	Leyland National 10351A/2R	B36D	1978	Ex Black Prince, Morley, 1993
1084	AKU161T	Leyland National 10351B/1R	B44F	1979	Ex Black Prince, Morley, 1993
1085	AVS238N	Leyland National 10351/1R/SC	DP39F	1975	Ex Sovereign, 1993
1086	BPL477T	Leyland National 10351B/1R	B44F	1979	Ex Sovereign, 1993
1087	AYR303T	Leyland National 10351A/2R	B36D	1978	Ex Ward, Market Harborough, 1993
1088	UHG718R	Leyland National 11351A/1R	B49F	1976	Ex Ribble, 1993
1089	SCK694P	Leyland National 11351A/1R	B49F	1976	Ex Ribble, 1993
1090	SCK696P	Leyland National 11351A/1R	B49F	1976	Ex Ribble, 1993
1091	ERP552T	Leyland National 11351A/1R	B49F	1979	Ex United Counties, 1993
1092	ERP553T	Leyland National 11351A/1R	B49F	1979	Ex United Counties, 1993
1093	ERP555T	Leyland National 11351A/1R	B49F	1979	Ex United Counties, 1993
1094	KRP564V	Leyland National 11351A/1R	B49F	1979	Ex United Counties, 1993
1095	MNH571V	Leyland National 11351A/1R	B49F	1979	Ex United Counties, 1993
1096	MNH576V	Leyland National 11351A/1R	B49F	1979	Ex United Counties, 1993
1097	PJT272R	Leyland National 11351A/1R	B49F	1977	Ex Quickstep, 1994
1098	PEV695R	Leyland National 11351A/1R	B49F	1976	Ex Quickstep, 1994

Livery: Red (buses); white and blue (coaches)

Previous Registrations:

AVS238N	HPF318N, HIL9374	CNB253M	PFM866M, 452WAL
BVU917N	JNA584N, 168WAL		

CHECKMATE

S M Tandy, 333 Stockport Road, Mossley, Ashton-u-Lyne,
Greater Manchester, OL5 0RS

E184CNE	Iveco Daily 49.10	Northern Counties	B22F	1988	Ex GM Buses, 1989
E975SVU	Freight Rover Sherpa	Made-to-Measure	C16F	1987	
F483JBU	Citroen C25D	Coachwork Walker	M14	1989	Ex private owner, 1990
F361FNB	Iveco Daily 49.10	Northern Counties	B22F	1988	

Livery: White and black

CITIBUS

Citibus Tours Ltd, Ward Street, Chadderton, Oldham, Greater Manchester, OL1 2HL

3-9 Mercedes-Benz 811D — Plaxton Beaver — B28F — 1992 — Ex Capital Citybus, 1994

3	J603HMF	6	J606HMF	7	J607HMF	8	J608HMF	9	J609HMF
4	J604HMF								

10	PRA10R	Leyland Leopard PSU3C/4R	Alexander AT	DP49F	1978	Ex Appleby, Conisholme, 1991
14	J614HMF	Mercedes-Benz 811D	Plaxton Beaver	B28F	1992	Ex Capital Citybus, 1994
19	J619HMF	Mercedes-Benz 811D	Plaxton Beaver	B28F	1992	Ex Capital Citybus, 1994
29	E329SWY	Iveco Daily 49.10	Robin Hood City Nippy	B23F	1988	Ex Keighley & District, 1993
46	E346SWY	Iveco Daily 49.10	Robin Hood City Nippy	B23F	1988	Ex Keighley & District, 1993
47	E447TYG	Iveco Daily 49.10	Robin Hood City Nippy	B23F	1988	Ex Keighley & District, 1993
50	E450TYG	Iveco Daily 49.10	Robin Hood City Nippy	DP25F	1988	Ex Keighley & District, 1993
51	E451TYG	Iveco Daily 49.10	Robin Hood City Nippy	DP19F	1988	Ex Country Bus & Coach, 1994
53	E453TYG	Iveco Daily 49.10	Robin Hood City Nippy	B23F	1988	Ex Keighley & District, 1993
56	E456TYG	Iveco Daily 49.10	Robin Hood City Nippy	B23F	1988	Ex Keighley & District, 1993
57	E457TYG	Iveco Daily 49.10	Robin Hood City Nippy	B23F	1988	Ex Keighley & District, 1993
74	FWA474V	Leyland National 2 NL106L11/1R		B44F	1980	Ex South Yorkshire's Transport, 1991
76	FWA476V	Leyland National 2 NL106L11/1R		B44F	1980	Ex South Yorkshire's Transport, 1991
85	PWY585W	Leyland National 2 NL106L11/1R		B44F	1980	Ex County Bus & Coach, 1993
86	PWY586W	Leyland National 2 NL106L11/1R		B44F	1980	Ex County Bus & Coach, 1993
104w	VNB104L	Leyland Atlantean AN68/1R	Park Royal	H43/32F	1972	Ex Isle of Man, 1988
604	PNW604W	Leyland National 2 NL116L11/1R		B52F	1980	Ex County Bus & Coach, 1993

The Oldham terminus of service 429 is now outside the shopping centre, and here Citibus 786, LPB185P is seen loading in readiness for its return to Manchester.
Phillip Stephenson

Now retired to Prestatyn on the North Wales coast, UDT189S, a Leyland Atlantean with East Lancashire bodywork was new to South Yorkshire. It was photographed while working service 135 to Bury Interchange. Several operators now compete along this route which is also served by Metrolink. *Phillip Stephenson*

601-658	Leyland Atlantean AN68A/1R	Roe	H45/29D*	1978-79	*615/7/8/29 are H45/32F Ex South Yorkshire's Transport, 1989-90

601	VET601S	**615**	VET615S	**622**	XWG622T	**633**	XWG633T	**641**	XWG641T
602	VET602S	**617**	VET617S	**625**	XWG625T	**637**	XWG637T	**642**	XWG642T
612w	VET612S	**618**	VET618S	**626**	XWG626T	**638**	XWG638T	**651**	XWG651T
613w	VET613S	**621**	XWG621T	**629**	XWG629T	**640**	XWG640T	**658**	XWG658T
614	VET614S								

672	YKY672T	Leyland Atlantean AN68A/1R	Roe	H45/32F	1979	Ex South Yorkshire's Transport, 1990
675	UET675S	Leyland Atlantean AN68A/1R	Alexander AL	H45/29D	1978	Ex South Yorkshire's Transport, 1991
676	MCA676T	Leyland National 10351B/1R		B44F	1979	Ex PMT, 1993
679	UET679S	Leyland Atlantean AN68A/1R	Alexander AL	H45/29D	1978	Ex South Yorkshire's Transport, 1991

689-732	Leyland Atlantean AN68A/1R	Alexander AL	H45/31F	1978-80	Ex South Yorkshire's Transport, 1991-92

689	CWG689V	**710**	CWG710V	**711**	CWG711V	**716**	CWG716V	**732**	CWG732V
690	CWG690V								

705	XEH254M	Leyland National 1051/1R/0402		B44F	1973	Ex Southend, 1993
734	CWF734T	Leyland Atlantean AN68A/1R	Roe	H45/29D	1979	Ex South Yorkshire's Transport, 1991
765	CWG765V	Leyland Atlantean AN68A/1R	Roe	H45/32F	1980	Ex South Yorkshire's Transport, 1991
774	CWG774V	Leyland Atlantean AN68A/1R	Roe	H45/32F	1980	Ex South Yorkshire's Transport, 1991
776	CWG776V	Leyland Atlantean AN68A/1R	Roe	H45/32F	1980	Ex South Yorkshire's Transport, 1991
782	GPD295N	Leyland National 10351/1R/SC		DP39F	1974	Ex Southend, 1993
783	HPF309N	Leyland National 10351/1R/SC		DP39F	1975	Ex Southend, 1993
784	HPF327N	Leyland National 10351/1R/SC		DP39F	1975	Ex Southend, 1993
785	HPF316N	Leyland National 10351/1R/SC		DP39F	1975	Ex Southend, 1993
786	LPB185P	Leyland National 10351/1R/SC		DP39F	1976	Ex Southend, 1993
787	LPB201P	Leyland National 10351/1R/SC		DP39F	1976	Ex Southend, 1993

Previous Registrations:
VNB104L VNB104L, 2705MAN

Livery: Blue and black.

CITY NIPPY

P V Walsh (J P Executive Travel), 121 Mount Road, Alkrington, Middleton, Manchester, M24 1EZ

D337VBB	Mercedes-Benz 307D	Mercedes-Benz	M12	1987	
E406EPE	Renault-Dodge S56	Northern Counties	B22F	1987	Ex Stagecoach South, 1993
E408EPE	Renault-Dodge S56	Northern Counties	B22F	1987	Ex Stagecoach South, 1993
E216WVM	Iveco Daily 49.10	Northern Counties	B22F	1988	Ex NCounties demonstrator, 1988
E357KPO	Iveco Daily 49.10	Robin Hood City Nippy	B25F	1988	
F726DNB	Iveco Daily 49.10	Robin Hood City Nippy	B25F	1988	
F638HVU	Renault-Dodge S56	Northern Counties	B22F	1988	
F639HVU	Renault-Dodge S56	Northern Counties	B22F	1988	
F377UCP	Mercedes-Benz 609D	Reeve Burgess Beaver	B24F	1988	Ex Aldham Coaches, Wombwell, 1994
F138KAO	Mercedes-Benz 609D	Reeve Burgess Beaver	B20F	1989	Ex North Western, 1994
G888TJA	Mercedes-Benz 709D	Phoenix	B29F	1990	
G889TJA	Mercedes-Benz 709D	Phoenix	B29F	1990	
H147CBU	Mercedes-Benz 709D	Phoenix	B29F	1991	
H691FNB	Mercedes-Benz 709D	Phoenix	B29F	1991	
H466HBA	Iveco Daily 49.10	Phoenix	B23F	1991	
J685PJA	Mercedes-Benz 609D	Made-to-Measure	B21F	1992	
J686PJA	Mercedes-Benz 609D	Made-to-Measure	B21F	1992	
J614PNE	Mercedes-Benz 609D	Made-to-Measure	B24F	1992	
J20JPT	Mercedes-Benz 609D	Made-to-Measure	B21F	1992	
K84UND	Mercedes-Benz 609D	Made-to-Measure	B21F	1992	
K123AJA	Mercedes-Benz 709D	Wright	B27F	1993	
L196DVM	Mercedes-Benz 709D	Wright	B27F	1993	
L502FVU	Mercedes-Benz 709D	Marshall C19	DP27F	1994	

Previous Registrations:
J20JPT J615PNE

Livery: Blue and silver

Seen at Middleton bus station bound for Alkrington is City Nippy H691FNB, a Mercedes-Benz 709D with Phoenix bodywork.

CUMBERLAND

Cumberland Motor Services, PO Box 17, Tangier Street, Whitehaven, Cumbria, CA28 7XF

Depots: Hindpool Road, Barrow; Willowholme Industrial Estate, Carlisle; Station Road, Kendal; Lillyhall; King Street, Millom; Brunswick Road, Penrith. **Outstations** Ambleside, Appleby, Askam, Grange, Haverthwaite, Millom, Orton, Penrith, Sedbergh and Ulverston.

34-46		Mercedes-Benz L608D		Reeve Burgess		B20F	1986-87		
34	D34UAO	**37**	D37UAO	**40**	D40UAO	**43**	D43UAO	**45**	D45UAO
35	D35UAO	**38**	D38UAO	**41**	D41UAO	**44**	D44UAO	**46**	D46UAO
36	D36UAO	**39**	D39UAO	**42**	D42UAO				

47-53		Mercedes-Benz 709D		Alexander AM		B25F	1988	51-53 Ex Hampshire Bus, 1989	
47	E47CHH	**49**	E49CHH	**51**	E510PVV	**52**	E511PVV	**53**	E512PVV
48	E48CHH	**50**	E50CHH						

54-70		Mercedes-Benz 709D		Alexander AM		B23F*	1990	55-70 ex Magicbus, 1990-91 *57-9,61-4 are B25F	
54	G178PAO	**58**	G268TSL	**61**	G263TSL	**64**	G266TSL	**67**	G295TSL
55	G299TSL	**59**	G269TSL	**62**	G264TSL	**65**	G297TSL	**68**	G294TSL
56	G300TSL	**60**	G296TSL	**63**	G265TSL	**66**	G298TSL	**70**	G293TSL
57	G267TSL								

71-78		Mercedes-Benz 709D		Alexander AM		B25F	1993		
71	K871GHH	**73**	K873GHH	**75**	K875GHH	**77**	K877GHH	**78**	K878GHH
72	K872GHH	**74**	K874GHH	**76**	K876GHH				

79-86		Mercedes-Benz 709D		Alexander AM		B25F	1993	Ex Ribble, 1994	
79	K626UFR	**81**	K622UFR	**83**	K121XHG	**85**	L123DRN	**86**	K113XHG
80	K623UFR	**82**	K114XHG	**84**	L126DRN				

90	E317BRM	MCW MetroRider MF150	MCW	C25F	1988	Ex Cook & Marshall, 1988
91	E855BRM	MCW MetroRider MF150	MCW	B25F	1988	Ex Cook & Marshall, 1988

Cumberland 124, J124AHH, is one their five Plaxton Expressliners. These have a special body based on Plaxton's Paramount 3500 III design to meet National Express' requirements. Appropriately, the vehicle is seen passing the Cumberland Hotel in Oxford Street, London. *Colin Lloyd*

101	109DRM	Leyland Tiger TRCTL11/3R	Duple Laser	C51F	1984	
102	A102DAO	Leyland Tiger TRCTL11/3R	Duple Laser	C50F	1984	
103	B103HAO	Leyland Tiger TRCTL11/3RH	Duple Laser 2	C50F	1984	
106	B106HAO	Leyland Tiger TRCTL11/3RH	Duple Laser 2	C53F	1984	
107	TCK841	Leyland Tiger TRCTL11/3RH	Duple Laser 2	C44FT	1985	
109	WLT706	Leyland Tiger TRCTL11/3RH	Plaxton Paramount 3500 II	C48FT	1987	
110	WLT824	Leyland Tiger TRCTL11/3RH	Plaxton Paramount 3500 II	C46FT	1987	
111	VRR447	Leyland Tiger TRCTL11/3RH	Plaxton Paramount 3500 II	C48FT	1985	Ex Hampshire Bus, 1988

120-124		Volvo B10M-60		Plaxton Expressliner		C46FT	1992		
120	J120AHH	**121**	J121AHH	**122**	J122AHH	**123**	J123AHH	**124**	J124AHH

125	L125NAO	Volvo B10M-60	Plaxton Expressliner 2	C46FT	1994	
126	L126NAO	Volvo B10M-60	Plaxton Expressliner 2	C46FT	1994	
127	L127NAO	Volvo B10M-60	Plaxton Expressliner 2	C46FT	1994	
149	IIL3503	Volvo B10M-61	Van Hool Alizée	C49FT	1988	Ex East Midland, 1993
150	IIL3505	Volvo B10M-61	Van Hool Alizée	C49FT	1988	Ex East Midland, 1993
151	VLF578	Volvo B10M-61	Van Hool Alizée	C48F	1981	Ex Magicbus, 1988
152	RUT842	Volvo B10M-61	Van Hool Alizée	C48F	1981	Ex Magicbus, 1988
153	LJC800	Volvo B10M-61	Van Hool Alizée	C48F	1982	Ex Magicbus, 1988
155	ORY640	Volvo B10M-61	Van Hool Alizée	C48F	1982	Ex Magicbus, 1988
156	PCK335	Leyland Tiger TRCTL11/3RH	Duple Laser 2	C53F	1985	Ex Ribble, 1989
157	WVT618	Volvo B10M-61	Plaxton Paramount 3500 III	C50F	1987	Ex Wallace Arnold, 1990
158	DSV943	Volvo B10M-61	Plaxton Paramount 3500 III	C50F	1987	Ex Wallace Arnold, 1990
161	JPU817	Volvo B10M-61	Plaxton Paramount 3500 III	C53F	1987	Ex Wallace Arnold, 1990
162	B162WRN	Leyland Tiger TRCTL11/3RH	Duple Laser 2	C53F	1985	Ex Ribble, 1991
251	F251JRM	Leyland Lynx LX112L10ZR1	Leyland	B51F	1989	
252	F252JRM	Leyland Lynx LX112L10ZR1	Leyland	B51F	1989	
253	F253KAO	Leyland Lynx LX112L10ZR1	Leyland	B51F	1989	
254	E709MFV	Leyland Lynx LX112L10ZR1	Leyland	B51F	1988	Ex Leyland Bus, 1989
255	C544RAO	Leyland Lynx LX1126LXCTFR1	Leyland	B51F	1986	Ex Ribble, 1991

270-282		Volvo B6-50		Alexander Dash		B40F	1993-94		
270	L270LHH	**272**	L272LHH	**274**	L274LHH	**276**	L276JAO	**282**	L282JAO
271	L271LHH	**273**	L273LHH	**275**	L275JAO				

420-437		Bristol VRT/SL3/6LXB		Eastern Coach Works		H43/31F	1980		
420	FAO420V	**424**	FAO424V	**428**	FAO428V	**432**	KRM432W	**435**	KRM435W
421	FAO421V	**425**	FAO425V	**429**	FAO429V	**433**	KRM433W	**436**	KRM436W
422	FAO422V	**426**	FAO426V	**430**	KRM430W	**434**	KRM434W	**437**	KRM437W
423	FAO423V	**427**	FAO427V	**431**	KRM431W				

Five B6Rs built by Volvo's subsidiary, Steyr, at its plant outside Vienna, Austria, were replaced by Scottish-built chassis in 1994. These took the same fleet numbers, and replacement 272, L272LHH, is seen in Carlisle shortly after delivery.
Brian Pritchard

5031	D503RCK	Mercedes-Benz L608D	Reeve Burgess	DP19F	1986	Ex Ribble, 1989
5041	D504RCK	Mercedes-Benz L608D	Reeve Burgess	DP19F	1986	Ex Ribble, 1989
505	LUA273V	Leyland Leopard PSU3F/4R	Plaxton Supreme IV	C53F	1980	Ex Yeowart, Whitehaven, 1988
509	E986AHH	DAF SB2305DHTD585	Plaxton Paramount 3200 III	C53F	1988	Ex Yeowart, Whitehaven, 1988
511	D511RCK	Mercedes-Benz L608D	Reeve Burgess	DP19F	1986	Ex Ribble, 1994
518	D518RCK	Mercedes-Benz L608D	Reeve Burgess	DP19F	1986	Ex Ribble, 1989
519	D519RCK	Mercedes-Benz L608D	Reeve Burgess	DP19F	1986	Ex Ribble, 1989
520	D520RCK	Mercedes-Benz L608D	Reeve Burgess	DP19F	1986	Ex Ribble, 1989

522-561

Mercedes-Benz L608D | Reeve Burgess | B20F | 1986 | Ex Ribble, 1989

522	D522RCK	**529**	D529RCK	**533**	D533RCK	**558**	D558RCK	**560**	D560RCK
525	D525RCK	**530**	D530RCK	**534**	D534RCK	**559**	D559RCK	**561**	D561RCK
528	D528RCK	**531**	D531RCK	**557**	D557RCK				

569	LUA275V	Leyland Leopard PSU3F/4R	Plaxton Supreme III	C53F	1980	Ex Kirkpatrick, Brigham, 1988
625	GRM625V	Leyland Leopard PSU3F/4R	Duple Dominant II	C49F	1980	

699-788

Volvo Citybus B10M-60 | Alexander PS | B49F* | 1992-93 | *772-791 are DP48F

699	K699ERM	**717**	K717DAO	**735**	K735DAO	**754**	K754DAO	**772**	K772DAO
700	K700DAO	**718**	K718DAO	**736**	K736DAO	**755**	K755DAO	**773**	K773DAO
701	K701DAO	**719**	K719DAO	**737**	K737DAO	**756**	K756DAO	**774**	K774DAO
702	K702DAO	**720**	K720DAO	**738**	K738DAO	**757**	K757DAO	**775**	K775DAO
703	K703DAO	**721**	K721DAO	**739**	K739DAO	**758**	K758DAO	**776**	K776DAO
704	K704ERM	**722**	K722DAO	**740**	K740DAO	**759**	K759DAO	**777**	K777DAO
705	K705DAO	**723**	K723DAO	**741**	K741DAO	**760**	K760DAO	**778**	K778DAO
706	K706DAO	**724**	K724DAO	**742**	K742DAO	**761**	K761DAO	**779**	K779DAO
707	K707DAO	**725**	K725DAO	**743**	K743DAO	**762**	K762DAO	**780**	K780DAO
708	K708DAO	**726**	K726DAO	**744**	K744DAO	**763**	K763DAO	**781**	K781DAO
709	K709DAO	**727**	K727DAO	**745**	K745DAO	**764**	K764DAO	**782**	K782DAO
710	K710DAO	**728**	K728DAO	**746**	K746DAO	**765**	K765DAO	**783**	K783DAO
711	K711DAO	**729**	K729DAO	**748**	K748DAO	**766**	K766DAO	**784**	K784DAO
712	K712DAO	**730**	K730DAO	**749**	K749DAO	**767**	K767DAO	**785**	K785DAO
713	K713DAO	**731**	K731DAO	**750**	K750DAO	**768**	K768DAO	**786**	K786DAO
714	K714DAO	**732**	K732DAO	**751**	K751DAO	**769**	K769DAO	**787**	K787DAO
715	K715DAO	**733**	K733DAO	**752**	K752DAO	**770**	K770DAO	**788**	K788DAO
716	K716DAO	**734**	K734DAO	**753**	K753DAO	**771**	K771DAO		

810	TRN810V	Leyland National 10351B/1R		B44F	1979	Ex Ribble, 1989
1001	URM801Y	Leyland Olympian ONLXB/1R	Eastern Coach Works	DPH45/30F	1982	
1002	URM802Y	Leyland Olympian ONLXB/1R	Eastern Coach Works	DPH45/30F	1982	

1003-1011

Leyland Olympian ONLXB/2RZ | Alexander RL | H51/36F | 1988

1003	F803FAO	**1005**	F805FAO	**1007**	F807FAO	**1009**	F809FAO	**1011**	F811FAO
1004	F804FAO	**1006**	F806FAO	**1008**	F808FAO	**1010**	F810FAO		

1012-1019

Leyland Olympian ON2R56G13Z4 Alexander RL | H51/34F | 1990

1012	H112SAO	**1014**	H114SAO	**1016**	H116SAO	**1018**	H118SAO	**1019**	H119SAO
1013	H113SAO	**1015**	H115SAO	**1017**	H117SAO				

1020-1027

Leyland Olympian ON2R56G13Z4 Alexander RL | DPH47/27F | 1991

1020	J120AAO	**1022**	J122AAO	**1024**	J124XHH	**1026**	J126XHH	**1027**	J127XHH
1021	J121AAO	**1023**	J123XHH	**1025**	J125XHH				

1028-1035

Leyland Olympian ON2R50G13Z4 Alexander RL | DPH43/27F | 1992

1028	K128DAO	**1030**	K130DAO	**1032**	K132DAO	**1034**	K134DAO	**1035**	K135DAO
1029	K129DAO	**1031**	K131DAO	**1033**	K133DAO				

1090	C382SAO	Leyland Olympian ONLXB/1RV	Alexander RL	H47/30F	1986	Ex Bluebird, 1991
1091	C383SAO	Leyland Olympian ONLXB/1RV	Alexander RL	H47/30F	1986	Ex Bluebird, 1991
1092	D384XAO	Leyland Olympian ONLXB/1RV	Alexander RL	H47/30F	1986	Ex Bluebird, 1991
1093	D380XRS	Leyland Olympian ONLXB/1RV	Alexander RL	H47/30F	1986	Ex Bluebird, 1992
1094	D381XRS	Leyland Olympian ONLXB/1RV	Alexander RL	H47/30F	1986	Ex Bluebird, 1992
1103	KRN103T	Leyland Leopard PSU3E/4R	Duple Dominant II	C47F	1980	Ex Ribble, 1986
1105	KRN105T	Leyland Leopard PSU3E/4R	Duple Dominant II	C47F	1980	Ex Ribble, 1986
1113	KRN113T	Leyland Leopard PSU3E/4R	Duple Dominant II	C47F	1980	Ex Ribble, 1986
1119	KRN119T	Leyland Leopard PSU3E/4R	Duple Dominant II	C47F	1980	Ex Ribble, 1986

First of the Alexander-bodied Volvo B10M-55s for the Stagecoach group went to Cumberland where they replaced all the Routemasters, and almost all the Leyland Nationals in a major investment programme. Photographed on service 70 in Carlyle is 701, K701DAO.
Roy Marshall

Cumberland provide many services for the tourist to the English Lake District, with vehicles painted in a green and cream scheme with Lakeland Experience lettering. Several are open-tops which ply between Bowness on the shore of Windermere, to Grasmere through Ambleside and Rydale. Seen leaving Windermere is 1928, ERV251D.
Roy Marshall

1140	WCK140V	Leyland Leopard PSU3E/4R	Duple Dominant II	C51F	1980	Ex Ribble, 1986
1151	B151WRN	Leyland Tiger TRCTL11/2R	Duple Laser 2	C49F	1985	Ex Ribble, 1991
1153	B153WRN	Leyland Tiger TRCTL11/2R	Duple Laser 2	C49F	1985	Ex Ribble, 1991
1154	B154WRN	Leyland Tiger TRCTL11/2R	Duple Laser 2	C49F	1985	Ex Ribble, 1991
1155	B43MAO	Leyland Tiger TRCTL11/3R	Duple Laser 2	C53F	1985	Ex Ribble, 1991
1162	WLT980	Volvo B10M-61	Plaxton Paramount 3500 II	C53F	1986	Ex Ribble, 1994
1175	MRJ275W	Leyland Leopard PSU5D/4R	Plaxton Supreme IV	C50F	1981	Ex Ribble, 1989
1199	FDV799V	Leyland Leopard PSU3E/4RT	Plaxton Supreme IV	C49F	1979	Ex Ribble, 1989
1201	F201FHH	Leyland Olympian ON6LXCT/3RZ	Alexander RL	DPH55/41F	1989	
1202	F202FHH	Leyland Olympian ON6LXCT/3RZ	Alexander RL	DPH55/41F	1989	
1253	HNE253V	Leyland Leopard PSU5C/4R	Duple Dominant II	C53F	1980	Ex Ribble, 1989
1928	ERV251D	Leyland Atlantean PDR1/1 MkII	Metro Cammell	O43/33F	1966	Ex Southdown, 1991
2002	CBV2S	Bristol VRT/SL3/501 (Gardner)	Eastern Coach Works	O43/31F	1977	Ex Ribble, 1986
2024	DBV24W	Bristol VRT/SL3/6LXB	Eastern Coach Works	H43/31F	1980	Ex Ribble, 1986
2032	DBV32W	Bristol VRT/SL3/6LXB	Eastern Coach Works	H43/31F	1980	Ex Ribble, 1986
2035	UWV610S	Bristol VRT/SL3/6LXB	Eastern Coach Works	O43/31F	1977	Ex Southdown, 1990
2036	UWV612S	Bristol VRT/SL3/6LXB	Eastern Coach Works	O43/31F	1977	Ex Southdown, 1990
2037	UWV618S	Bristol VRT/SL3/6LXB	Eastern Coach Works	O43/31F	1978	Ex Southdown, 1990
2038	UWV620S	Bristol VRT/SL3/6LXB	Eastern Coach Works	O43/31F	1978	Ex Southdown, 1990

Leyland Olympians, with Alexander RL-type bodywork and high-back seating are the nomal vehicle to be found on the 555 service between Lancaster and Keswick. Advertised as one of the most delightful stage services, the route passes several of the lakes, including Windermere, Rydal Water and Derwent Water. Photographed heading south is 1021, J121AAO. *Roy Marshall*

2134	DBV134Y	Leyland Olympian ONLXB/1R	Eastern Coach Works	H45/32F	1983	Ex Ribble, 1989
2175	C175ECK	Leyland Olympian ONLXB/1R	Eastern Coach Works	DPH41/26F	1985	Ex Ribble, 1989
2176	C176ECK	Leyland Olympian ONLXB/1R	Eastern Coach Works	DPH41/26F	1985	Ex Ribble, 1989
2177	C177ECK	Leyland Olympian ONLXB/1R	Eastern Coach Works	DPH41/26F	1985	Ex Ribble, 1989

Previous Registrations:

109DRM	A101DAO	LJC800	From new
B43MAO	B155WRN, PCK335	ORY640	From new
C382SAO	C473SSO, GSO3V	PCK335	B156WRN
C383SAO	C474SSO, GSO4V	RUT842	From new
D384XAO	D375XRS, GSO5V	TCK841	B107HAO
D560RCK	D561RCK	VLF578	TGD766W
D561RCK	D560RCK	VRR447	B180RLJ
DSV943	D203LWX	WLT706	C109OHH
IIL3503	E625UNE, TXI2426, E936XSB	WLT824	C110OHH
IIL3505	E623UNE, XIA257, E942XSB	WLT980	C105DWR
JPU817	D207LWX	WVT618	D202LWX

Livery : Stagecoach white, red, orange and blue

Coachline:	90, 102-9/11, 151-3/5/6/62, 509/69,
Lakeland Experience:	520/58/60, 810, 1928, 2002/35-8.
National Express:	110, 120-7/49/50
Shearings Holidays:	101/57/8/61.

Named vehicles:
520 *William Wordsworth*, 558 *John Ruskin*, 560 *Beatrix Potter.*

DALYBUS

Dalybus Ltd, 89 Barton Lane, Eccles, Manchester, M30 0EY

36	E36VKP	Iveco Daily 49.10	Dormobile Routemaker	B25F	1987	Ex A D Coaches, Dundonald, 1993
80	E80PUH	Freight Rover Sherpa	Carlyle Citybus 2	B20F	1987	Ex Shamrock, Pontypridd, 1994
94	E94OUH	Freight Rover Sherpa	Carlyle Citybus 2	B20F	1987	Ex Red & White, 1994
107	E107OUH	Freight Rover Sherpa	Carlyle Citybus 2	B20F	1987	Ex Red & White, 1994
145	D145WCC	Freight Rover Sherpa	Carlyle Citybus 2	B20F	1987	Ex Midland, 1993
146	E146RNY	Freight Rover Sherpa	Carlyle Citybus 2	B20F	1987	Ex Red & White, 1994
147	D147WCC	Freight Rover Sherpa	Carlyle Citybus 2	B20F	1987	Ex Morris, Mold, 1993
151	E151RNY	Freight Rover Sherpa	Carlyle Citybus 2	B20F	1987	Ex Red & White, 1993
152	E152RNY	Freight Rover Sherpa	Carlyle Citybus 2	B20F	1987	Ex Red & White, 1993
154	E154RNY	Freight Rover Sherpa	Carlyle Citybus 2	B18F	1988	Ex Red & White, 1994
185	PKE185W	Bedford YMT	Duple Dominant II Express	C45F	1981	Ex Boro'line, 1992
211	F211AKG	Freight Rover Sherpa	Carlyle Citybus 2	B20F	1988	Ex Shamrock, Pontypridd, 1994
214	F214AKG	Freight Rover Sherpa	Carlyle Citybus 2	B20F	1988	Ex Phil Anslow, Garndiffaith, 1993
229	F229BAX	Freight Rover Sherpa	Carlyle Citybus 2	B18F	1989	Ex Red & White, 1994
281	G281HDW	Freight Rover Sherpa	Carlyle Citybus 2	B20F	1990	Ex National Welsh, 1992
328	JMB328T	Leyland Leopard PSU3E/4R	Duple Dominant II Express	DP55F	1978	Ex Midland, 1994
721	H721LOL	Freight Rover Sherpa	Carlyle Citybus 2	B20F	1990	Ex Merry Hill Mini, 1993
725	H725LOL	Freight Rover Sherpa	Carlyle Citybus 2	B20F	1990	Ex Merry Hill Mini, 1993
807	C807TLF	Bedford YMPS	Plaxton Paramount 3200 II	C35F	1986	Ex Capital, West Drayton, 1993

Livery: Yellow, red, white and blue.

DARWEN COACH SERVICES

D R Russell, 5 West View, Chapels, Darwen, Lancashire, BB3 0NP

B931YCW	Mercedes-Benz L307D	Rovanne	M12	1985	Ex ?, 1987
D711HUA	Freight Rover Sherpa	Optare	B16F	1986	Ex Fishwick, 1989
D712HUA	Freight Rover Sherpa	Optare	B16F	1986	Ex Fishwick, 1989
D181NON	Freight Rover Sherpa	Carlyle	B20F	1987	Ex Bee Line Buzz, 1992
F217AKG	Freight Rover Sherpa	Carlyle Citybus 2	B20F	1988	Ex National Welsh, 1989
F948CUA	Freight Rover Sherpa	Carlyle Citybus 2	B20F	1988	Ex Yorkshire Rider, 1993

Livery: Green

DAVID TANNER TRAVEL

D Tanner, 10 Cowley Street, Haydock, St Helens, Merseyside, WA10 2SR

KRE278P	Leyland National 11351/1R		B52F	1976	Ex PMT, 1992
B930KWM	Quest B	Locomotors	DP23F	1985	Ex Banga, Wolverhampton, 1994
D109OWG	Renault-Dodge S56	Reeve Burgess	B25F	1987	Ex Somerbus, Paulton, 1994

Blackburn rail station is the background to this picture of Darwen Coach Services F948CUA. New to Yorkshire Rider, this fleet now operate in a similar scheme to two similar minibuses obtained from Fishwick in 1989 *Andrew Jorosz*

KRE278P was one of the later Leyland Nationals which was built with the long, lower pod to the order of PMT, whose basic livery the vehicle retains under its new operation with Davis Tanner Travel. *Andrew Jarosz*

DENNIS'S

R & M Cooper, 217 Stamford Street Central, Ashton-under-Lyne, Greater Manchester, OL6 7QB

NEV682M	Leyland National 1151/1R/0402		B49F	1973	Ex Edinburgh Transport, 1994
TOF717S	Leyland National 11351/1R		B49F	1977	Ex Midland, 1994
D535XBC	Mercedes-Benz 709D	Advanced Vehicle Builders	DP25F	1987	
E996VYS	Mercedes-Benz 609D	Reeve Burgess Beaver	C19F	1987	Ex Henderson, Hamilton, 1993
F220JBB	Mercedes-Benz 609D	Reeve Burgess Beaver	C21F	1988	
G506PFM	Mercedes-Benz 811D	PMT	B29F	1989	
G507PFM	Mercedes-Benz 811D	PMT	B29F	1989	
H210CVU	Mercedes-Benz 811D	Reeve Burgess Beaver	B33F	1990	
H211CVU	Mercedes-Benz 814D	Reeve Burgess Beaver	C29F	1990	
H933DBU	Mercedes-Benz 811D	Reeve Burgess Beaver	B33F	1990	
J23GCX	DAF SB220LC550	Optare Delta	B49F	1991	
K286ESF	Mercedes-Benz 709D	Dormobile Routemaker	B27F	1993	
K287ESF	Mercedes-Benz 709D	Dormobile Routemaker	B27F	1993	
K289ESF	Mercedes-Benz 709D	Dormobile Routemaker	B27F	1993	
K290ESF	Mercedes-Benz 709D	Dormobile Routemaker	B27F	1993	
L680GNA	Mercedes-Benz 709D	Plaxton Beaver	B27F	1994	
L681GNA	Mercedes-Benz 709D	Plaxton Beaver	B27F	1994	
L682GNA	Mercedes-Benz 709D	Plaxton Beaver	B27F	1994	
L683GNA	Mercedes-Benz 709D	Plaxton Beaver	B27F	1994	
L510JND	Mercedes-Benz 709D	Plaxton Beaver	B27F	1994	
M728MBU	Mercedes-Benz 709D	Plaxton Beaver	B27F	1994	
M729MBU	Mercedes-Benz 709D	Plaxton Beaver	B27F	1994	
M730MBU	Mercedes-Benz 709D	Plaxton Beaver	B27F	1994	

Livery: Red and grey

EXPRESS TRAVEL

Express Travel Ltd, Woodend Avenue, Speke, Liverpool, L24 9NB

H842AHS	Volvo B10M-60	Plaxton Paramount 3500 III	C49FT	1991	Ex Park's, Hamilton, 1992
H843AHS	Volvo B10M-60	Plaxton Paramount 3500 III	C49FT	1991	Ex Park's, Hamilton, 1992
H844AHS	Volvo B10M-60	Plaxton Paramount 3500 III	C49FT	1991	Ex Park's, Hamilton, 1992
H845AHS	Volvo B10M-60	Plaxton Paramount 3500 III	C49FT	1991	Ex Park's, Hamilton, 1992
J246MFP	Volvo B10M-60	Plaxton Paramount 3500 III	C49FT	1992	Ex Tellings-Golden Miller, Cardiff, 1993
J247MFP	Volvo B10M-60	Plaxton Paramount 3500 III	C46FT	1992	Ex Tellings-Golden Miller, Cardiff, 1993
J16AMB	DAF SB3000DKVF601	Van Hool Alizée H	C46FT	1992	
J17AMB	DAF SB3000DKVF601	Van Hool Alizée H	C46FT	1992	
K18AMB	Volvo B10M-60	Plaxton Expressliner II	C46FT	1992	
K19AMB	Volvo B10M-60	Plaxton Expressliner II	C46FT	1992	
K20AMB	Volvo B10M-60	Plaxton Expressliner II	C46FT	1992	
K22AMB	DAF SB3000DKVF601	Van Hool Alizée H	C46FT	1993	
K504WNR	Volvo B10M-60	Plaxton Expressliner II	C46FT	1993	
K505WNR	Volvo B10M-60	Plaxton Expressliner II	C46FT	1993	
K506WNR	Volvo B10M-60	Plaxton Expressliner II	C46FT	1993	
L705PHE	Volvo B10M-62	Van Hool Alizée	C46FT	1994	
L706PHE	Volvo B10M-62	Van Hool Alizée	C46FT	1994	
L707PHE	Volvo B10M-62	Van Hool Alizée	C46FT	1994	
L708PHE	Volvo B10M-62	Van Hool Alizée	C46FT	1994	
L709PHE	Volvo B10M-62	Van Hool Alizée	C46FT	1994	
L710PHE	Volvo B10M-62	Van Hool Alizée	C46FT	1994	
L711PHE	Volvo B10M-62	Van Hool Alizée	C46FT	1994	
L712PHE	Volvo B10M-62	Van Hool Alizée	C46FT	1994	
L713PHE	Volvo B10M-62	Van Hool Alizée	C46FT	1994	
L714PHE	Volvo B10M-62	Van Hool Alizée	C46FT	1994	

Livery:

Dennis's operate mostly Mercedes-Benz minibuses on their services to the east of Manchester. However, three full-size single-decks are used, mostly on service 216 between Manchester and Ashton-under-Lyne. Photographed in Manchester's Piccadilly is J23GCX, a DAF SB220LC550 with Optare Delta bodywork

Express Travel are shortly to commence a quality express service between Liverpool, Manchester and London. The initial fleet, as shown aside, has come from the former Express Travel operations in Perth and speke which were part of British Bus plc, though Express Travel itself is no longer part of that group. *Andrew Jarosz*

FINGLANDS

Finglands Coachways Ltd, 261 Wilmslow Road, Rusholme, Manchester, M14 5LJ

A subsidiary of EYMS Holdings Ltd

52	152FRH	Leyland Tiger TRCTL11/3RZ	Duple 340	C51FT	1988	Ex East Yorkshire, 1993
301	F301JNC	Leyland Tiger TRBTL11/2RZ	Duple 300	B55F	1989	
302	F302JNC	Leyland Tiger TRBTL11/2RZ	Duple 300	B55F	1989	
343	F303JNC	Mercedes-Benz 609D	PMT	DP24F	1989	Ex Fieldsend, Rusholme, 1992
344	HIL7748	Mercedes-Benz 609D	Reeve Burgess Beaver	DP19F	1990	
351	JNB151N	Leyland Leopard PSU3C/4R	Plaxton Elite III Express	C53F	1975	
352	CIB3202	Leyland Tiger TRCTL11/3R	Plaxton Paramount 3200	C57F	1984	
355	HIL7747	Leyland Tiger TRCTL11/3RZ	Plaxton Paramount 3200 II	C57F	1985	
356	HIL7923	Leyland Tiger TRCTL11/3RZ	Plaxton Paramount 3200 II	C57F	1985	Ex Fieldsend, Rusholme, 1992
357	RYV77	Volvo B10M-61	Caetano Algarve	C53F	1986	
358	HIL7745	Volvo B10M-61	Plaxton Paramount 3500 III	C28FT	1986	
359	HIL7746	Volvo B10M-61	Plaxton Paramount 3500 III	C53FT	1989	
360	10RU	Mercedes-Benz 0303/15R	Mercedes-Benz	C49FT	1990	
361	SIA6180	Mercedes-Benz 0303/15R	Mercedes-Benz	C53FT	1990	
362	647JOE	Leyland Tiger TRCL10/3ARZM	Plaxton Paramount 3500 III	C49FT	1991	
	H907AHS	Volvo B10M-60	Plaxton Paramount 3500 III	C53F	1991	Ex Park's, Hamilton, 1993
903	UNA844S	Leyland Atlantean AN68A/1R	Northern Counties	H43/32F	1974	Ex East Yorkshire, 1993
904	LJA611P	Leyland Atlantean AN68/1R	Northern Counties	H43/32F	1974	Ex East Yorkshire, 1993
934	SRJ734R	Leyland Atlantean AN68A/1R	Northern Counties	H41/32F	1977	Ex East Yorkshire, 1993
937	BNC947T	Leyland Atlantean AN68A/1R	Park Royal	H43/32F	1979	Ex East Yorkshire, 1993
938	UNA848S	Leyland Atlantean AN68A/1R	Park Royal	H43/32F	1978	Ex East Yorkshire, 1993
939	UNA819S	Leyland Atlantean AN68A/1R	Park Royal	H43/32F	1978	Ex East Yorkshire, 1993
1707	YNA297M	Daimler Fleetline CRG6LXB	Northern Counties	H43/32F	1973	Ex Greater Manchester, 1988
1710	F242MBA	Volvo Citybus B10M-50	Alexander RV	DPH43/35F	1989	
1714	VNB157L	Leyland Atlantean AN68/1R	Park Royal	H43/32F	1972	Ex Impact, Old Trafford, 1989
1716	VNB173L	Daimler Fleetline CRG6LXB	Northern Counties	H43/32F	1973	Ex Impact, Old Trafford, 1989
1718	XJA515L	Daimler Fleetline CRG6LXB	Park Royal	H43/32F	1973	Ex Arlington, 1990
1719	WBN983L	Daimler Fleetline CRG6LXB	Northern Counties	H43/32F	1973	Ex preservation, 1990
1720	ONF695R	Leyland Atlantean AN68A/1R	Northern Counties	H43/32F	1977	Ex East Kent, 1991
1721	ONF653R	Leyland Atlantean AN68A/1R	Northern Counties	H43/32F	1977	Ex East Kent, 1991
1722	ONF663R	Leyland Atlantean AN68A/1R	Northern Counties	H43/32F	1977	Ex East Kent, 1991
1723	RJA703R	Leyland Atlantean AN68A/1R	Northern Counties	H43/32F	1977	Ex East Kent, 1991

Differences between the Northern Counties and Park Royal designs of standard bodies can be seen in this picture of Finglands 904, LJA611P and 937, BNC947T. Both the vehicles were new to Greater Manchester, arriving at Finglands from East Yorkshire's main fleet.
Andrew Jarosz

Five MCW Metrobus DR102s now ply Wilmslow Road in Finglands livery, all new to London Buses in 1987. Photographed at the city end of service 42 is 1729, E477SON. *Phillip Stephenson*

1726	E473SON	MCW Metrobus DR102/63	MCW	H45/30F	1987	Ex London Buses, 1992
1727	E474SON	MCW Metrobus DR102/63	MCW	H45/30F	1987	Ex London Buses, 1992
1728	E476SON	MCW Metrobus DR102/63	MCW	H45/30F	1987	Ex London Buses, 1992
1729	E477SON	MCW Metrobus DR102/63	MCW	H45/30F	1987	Ex London Buses, 1992
1730	E480UOF	MCW Metrobus DR102/65	MCW	H45/30F	1987	Ex London Buses, 1992
1731	LTK91R	Leyland Atlantean AN68A/1R	Roe	H43/30F	1977	Ex Hyndburn, 1992
1732	LTK93R	Leyland Atlantean AN68A/1R	Roe	H43/30F	1977	Ex Hyndburn, 1992
1733	LTK94R	Leyland Atlantean AN68A/1R	Roe	H43/30F	1977	Ex Hyndburn, 1992
1734	LTK96R	Leyland Atlantean AN68A/1R	Roe	H43/30F	1977	Ex Hyndburn, 1992
1735	NRN397P	Leyland Atlantean AN68/1R	Park Royal	H43/30F	1976	Ex East Yorkshire, 1993
1736	JDB121N	Leyland Atlantean AN68A/1R	Northern Counties	H43/32F	1975	Ex Ribble, 1993
1737	KBU912P	Leyland Atlantean AN68A/1R	Northern Counties	H43/32F	1975	Ex Ribble, 1993
1738	KDB686P	Leyland Atlantean AN68A/1R	Northern Counties	H43/32F	1975	Ex Ribble, 1993

Previous Registrations:

10RU	From new	CIB3202	A402HRJ	HIL7748	G880VVR
152FRH	E52URH	HIL7745	D400MND	HIL7923	B106WRJ
39EYD	C115DRH	HIL7746	F300JNC	RYV77	C666DJA
647JOE	From new	HIL7747	B88WRJ	SIA6180	From new

Livery: White, orange and brown.

Fishwicks traditional services between Preston, Leyland and Chorley continue almost unchanged. The vehicle fleet has, however, moved forward, with Leyland Lynx replacing early Leyland Nationals, and Volvo B10Ms replacing the Leyland Tiger coaches. 1993/94 has seen the arrival of four second-hand double decks to work school contracts. Representing the fleet is *(top)* 8, GCK429W, a Leyland National 2 seen in Preston; 30, D30VCW, a Leyland Lynx photographed in Chorley on service 109 and new coach L1JFS, an EOS90 integral coach. The latter making an appearance at the 1994 Ribble Enthusiasts' Club rally between Blackpool and Southport.
Paul Wigan/Phillip Stephenson/Simon Watts

FISHWICK

John Fishwick & Sons, Golden Hill Garage, Golden Hill Lane, Leyland, Lancashire, PR5 2LE

1	NFR558T	Leyland National 11351A/1R		B49F	1979	
2	A462LFV	Leyland Atlantean AN69/2L	Eastern Coach Works	H47/35F	1983	
3	J7JFS	Leyland Lynx LX2R11C15Z4R	Leyland Lynx 2	B51F	1991	
4	H64CCK	Leyland Lynx LX2R11C15Z4R	Leyland Lynx 2	B47F	1991	
5	H65CCK	Leyland Lynx LX2R11C15Z4R	Leyland Lynx 2	B51F	1991	
6	NFR559T	Leyland National 11351A/1R		B49F	1979	
7	GCK428W	Leyland National 2 NL116AL11/1R		B49F	1981	
8	GCK429W	Leyland National 2 NL116AL11/1R		B49F	1981	
9	WRN412V	Leyland National 2 NL116L11/1R		B49F	1980	
10	WRN413V	Leyland National 2 NL116L11/1R		B49F	1980	
11w	NFR560T	Leyland National 11351A/1R		B49F	1979	
12	GCK430W	Leyland National 2 NL116AL11/1R		B49F	1981	
14	J14JFS	Leyland Lynx LX2R11C15Z4R	Leyland Lynx 2	B51F	1992	
15	UMR194T	Leyland Fleetline FE30AGR	Eastern Coach Works	H43/31F	1979	Ex Thamesdown, 1993
16	OFV620X	Leyland National 2 NL116AL11/1R		B49F	1981	
17	UMR196T	Leyland Fleetline FE30AGR	Eastern Coach Works	H43/31F	1979	Ex Thamesdown, 1994
18	XTB728N	Leyland Atlantean AN68/1R	East Lancashire	H43/31F	1974	
19	XTB729N	Leyland Atlantean AN68/1R	East Lancashire	H43/31F	1974	
20	LUA714V	Bristol VRT/SL3/6LXB	Eastern Coach Works	H43/31F	1979	Ex ?,1994
21	DWU295T	Bristol VRT/SL3/6LXB	Eastern Coach Works	H43/31F	1979	Ex ?,1994
22	SRN103P	Leyland Atlantean AN68A/1R	East Lancashire	H43/32F	1976	
23	GRN895W	Leyland Atlantean AN69/1L	Eastern Coach Works	H43/31F	1981	
24	XCW955R	Leyland National 11351A/1R		B49F	1977	
25	D25VCW	Leyland Lynx LX112TL11ZR1	Leyland	B47F	1986	
26	OFV621X	Leyland National 2 NL116AL11/1R		B49F	1981	
27	ABV939Y	Leyland National 2 NL116TL11/1R		B49F	1982	
28	FBV524S	Leyland National 11351A/1R		B49F	1978	
30	D30VCW	Leyland Lynx LX112TL11ZR1	Leyland	B47F	1986	
32	D32YCW	Leyland Lynx LX112TL11ZR1	Leyland	B47F	1987	
33	D33YCW	Leyland Lynx LX112TL11ZR1	Leyland	B47F	1987	
34w	XCW957R	Leyland National 11351A/1R		B49F	1977	
C1	J9JFS	DAF SB3000DKV601	Van Hool Alizée	C49FT	1991	
C2	G22MHG	Volvo B10M-60	Van Hool Alizée	C49FT	1988	
C3	G23MHG	Volvo B10M-60	Van Hool Alizée	C49FT	1988	
C4	K3JFS	Volvo B10M-60	Plaxton Premiére 350	C49FT	1993	
C5	K4JFS	Volvo B10M-60	Plaxton Premiére 350	C50F	1993	
C6	G53RGG	Volvo B10M-60	Plaxton Paramount 3500 III	C53F	1988	Ex Park's, Hamilton, 1992
C7	G54RGG	Volvo B10M-60	Plaxton Paramount 3500 III	C53F	1988	Ex Park's, Hamilton, 1992
C8	G55RGG	Volvo B10M-60	Plaxton Paramount 3500 III	C53F	1988	Ex Park's, Hamilton, 1992
C9	G56RGG	Volvo B10M-60	Plaxton Paramount 3500 III	C53F	1988	Ex Park's, Hamilton, 1992
L1	L1JFS	EOS 90	EOS	C48FT	1994	
M1	F705WFV	Mercedes-Benz 609D	Reeve Burgess Beaver	C19F	1989	
M2	F706WFV	Mercedes-Benz 609D	Reeve Burgess Beaver	C19F	1989	
M5	E45HBV	Mercedes-Benz 609D	Reeve Burgess Beaver	B20F	1987	
M6	E46HBV	Mercedes-Benz 609D	Reeve Burgess Beaver	B20F	1987	
M7	E47HBV	Mercedes-Benz 609D	Reeve Burgess Beaver	B20F	1987	
M8	D84BLF	Mercedes-Benz 609D	Reeve Burgess	C19F	1987	Ex Mercedes-Benz, 1987
M9	D210OKY	Mercedes-Benz 709D	Reeve Burgess	B20F	1986	Ex Reeve Burgess, 1988
M10	E100MFV	Mercedes-Benz 609D	Reeve Burgess Beaver	C19F	1988	
M11	K5JFS	Mercedes-Benz 814D	Autobus Classique 2	C29F	1993	

Livery: Two-tone green.

Previous Registrations:

K5JFS	K721GBE

GLOSSOPDALE

D Whyatt, Glossopdale Bus Company, 5 Railway Street, Glossop, Derbyshire, SK13 9AG

B118RRE	Ford Transit 160	PMT	B16F	1985	Ex Lancaster, 1992
C203GKR	Bedford YMT	Wright TT	DP53F	1986	Ex Boro'line, 1992
D110OWG	Renault-Dodge S56	Reeve Burgess Beaver	B25F	1986	Ex South Yorkshire's Transport, 1991
D117OWG	Renault-Dodge S56	Reeve Burgess Beaver	B25F	1986	Ex Stevensons, 1992
D122OWG	Renault-Dodge S56	Reeve Burgess Beaver	B25F	1987	Ex Stevensons, 1992
D864LWR	Freight Rover Sherpa	Dormobile	B20F	1987	Ex Yorkshire Rider, 1991
E84OUH	Freight Rover Sherpa	Carlyle Citybus 2	B20F	1988	Ex Fen Travel, Syston, 1993
E157RNY	Freight Rover Sherpa	Carlyle Citybus 2	B20F	1988	Ex Red & White, 1994
E176TWO	Freight Rover Sherpa	Carlyle Citybus 2	B20F	1988	Ex West London, Tylers Green, 1994
M634FJF	Mercedes-Benz 811D	Marshall C16	B31F	1994	
M635FJF	Mercedes-Benz 711D	Marshall C19	B27F	1994	
M636FJF	Mercedes-Benz 711D	Marshall C19	B27F	1994	

Livery: Green and yellow

New arrivals at Glossopdale are three Mercedes-Benz, all with Marshall bodywork. One is the longer 811 version while two, including M636FJF, seen here in Stockport bus station, are of the 711 type. *Andy Jarosz*

GMN

Greater Manchester Buses North Ltd, Wallshaw Place,
Oldham, Greater Manchester, OL1 3UY

Depots: Howe Bridge, Atherton; Crook Street, Bolton; Rochdale Road, Bury; Boyle Street, Manchester; Queens Road, Manchester; Mumps Bridge, Oldham; Southport and Melverley Street, Wigan.

132	JNA588N	Leyland National 10351/1R				B41F	1975		
179-247		Leyland National 11351A/1R				B49F	1977/79		
179	RBU179R	**201**	ABA26T	**202**	ABA27T	**203**	ABA28T	**247**	WBN481T
197	ABA22T								
436-463		Leyland Leopard PSU3D/2R		Plaxton Derwent		B48F	1976		
436	LTE487P	**441**	LTE492P	**452**	MTE20R	**457**	MTE25R	**461**	MTE29R
438	LTE489P	**442**	LTE493P	**454**	MTE22R	**458**	MTE26R	**462**	MTE30R
439	LTE490P	**444**	LTE495P	**455**	MTE23R	**459**	MTE27R	**463**	MTE31R
440	LTE491P	**451**	MTE19R	**456**	MTE24R	**460**	MTE28R		
501	D501LNA	Leyland Lynx LX563LXCTZR1		Leyland		B48F	1986		
502	D502LNA	Leyland Lynx LX563LXCTZR1		Leyland		B48F	1986		
503	D503LNA	Leyland Lynx LX563LXCTZR1		Leyland		B48F	1986		
504	D504LNA	Leyland Lynx LX563LXCTZR1		Leyland		B48F	1986		
701	J461OVU	Volvo Citybus B10M-55		Northern Counties Paladin		B49F	1991		

The Northern Counties Paladin used to launch the type was GM Buses 701, J461OVU, now allocated to GMN at Wigan depot and photographed in its home town. The vehicle remained the newest single deck bus until GMN's recent delivery of B6s that carry the Mark 2 version of the Paladin.
Andrew Jarosz

Three Volvo Citybus double decks joined the evaluation fleet in 1986, with ten more being delivered to the main fleet in 1991-92. All moved to GMN, the original three being based at Wigan where 1481, C481CBU was photogrphed. *Phillip Stephenson*

GMN have gathered MCW MetroRiders into the fleet, some being allocated to Southport for services on Merseyside. Examples from Cardiff, SWT, London Buses and Yorkshire Rider have entered service. Photographed in Bolton's Moor Lane bus station is 1686, E636KYW, one of the former-London Buses examples. *Andrew Jarosz*

1051-1060	Volvo B6-50	Northern Counties Paladin 2	B40F	1994	

1051	M251NVM	**1053**	M253NVM	**1055**	M255NVM	**1057**	M257NVM	**1059**	M259NVM
1052	M252NVM	**1054**	M254NVM	**1056**	M256NVM	**1058**	M258NVM	**1060**	M260NVM

1481	C481CBU	Volvo Citybus B10M-50	Northern Counties	H46/33F	1986
1482	C482CBU	Volvo Citybus B10M-50	Northern Counties	H46/33F	1986
1483	C483CBU	Volvo Citybus B10M-50	Northern Counties	H46/33F	1986

1575-1580	MCW MetroRider MF150/30	MCW	B23F	1987	Ex Cardiff, 1994

1575	E120RDW	**1577**	E122RDW	**1578**	E123RDW	**1579**	E124RDW	**1580**	E125RDW
1576	E121RDW								

1581-1599	MCW MetroRider MF150/16	MCW	B25F	1987	Ex SWT, 1994

1581	E254REP	**1585**	E259REP	**1589**	E265REP	**1593**	E271REP	**1597**	E275REP
1582	E255REP	**1586**	D260PEP	**1590**	E266REP	**1594**	E272REP	**1598**	E276REP
1583	D256PEP	**1587**	E263REP	**1591**	E267REP	**1595**	E273REP	**1599**	E277REP
1584	E258REP	**1588**	E264REP	**1592**	E268REP	**1596**	E274REP		

1616	D616MDB	MCW MetroRider MF151/3	MCW	B23F	1987	
1622	D622MDB	MCW MetroRider MF151/3	MCW	B23F	1987	
1625	D625MDB	MCW MetroRider MF151/3	MCW	B23F	1987	
1631	D631MDB	MCW MetroRider MF151/3	MCW	B23F	1987	
1681	D474PON	MCW MetroRider MF150/14	MCW	B23F	1987	Ex London Buses, 1993
1682	E148KYW	MCW MetroRider MF150/38	MCW	B25F	1987	Ex London Buses, 1993
1683	E150KYW	MCW MetroRider MF150/38	MCW	B25F	1987	Ex London Buses, 1993
1684	E638KYW	MCW MetroRider MF150/46	MCW	B25F	1988	Ex London Buses, 1993
1685	E637KYW	MCW MetroRider MF150/46	MCW	B25F	1988	Ex London Buses, 1993
1686	E636KYW	MCW MetroRider MF150/46	MCW	B25F	1988	Ex London Buses, 1993
1687	E929KYR	MCW MetroRider MF150/46	MCW	B25F	1988	Ex London Buses, 1993

1688-1698	MCW MetroRider MF150/80	MCW	B23F	1988	Ex Yorkshire Rider, 1994

1688	E237UWR	**1691**	E241UWR	**1693**	E243UWR	**1695**	E246UWR	**1697**	E250UWR
1689	E239UWR	**1692**	E242UWR	**1694**	E244UWR	**1696**	E247UWR	**1698**	E251UWR
1690	E240UWR								

1722	H722CNC	Renault S75	Northern Counties	B17FL	1990
1723	H723CNC	Renault S75	Northern Counties	B17FL	1990
1725	H725CNC	Renault S75	Northern Counties	B17FL	1990

1777-1799	Renault-Dodge S56	Northern Counties	B20F	1987

1777	D777RBU	**1783**	E783SJA	**1785**	E785SJA	**1790**	E790SJA	**1796**	E796SJA
1778	D778RBU	**1784**	E784SJA	**1786**	E786SJA	**1794**	E794SJA	**1799**	E799SJA
1780	D780RBU								

1845-1899	Renault-Dodge S56	Northern Counties	B20F	1986-87

1845	D845LND	**1854**	D854LND	**1871**	D871MDB	**1879**	D879MDB	**1894**	D894MDB
1846	D846LND	**1857**	D857LND	**1874**	D874MDB	**1885**	D885MDB	**1897**	D897MDB
1847	D847LND	**1864**	D864LND	**1875**	D875MDB	**1889**	D889MDB	**1899**	D899MDB
1852	D852LND	**1870**	D870MDB	**1876**	D876MDB				

1904-1997	Renault-Dodge S56	Northern Counties	B20F	1987

1904	D904MDB	**1928**	D928NDB	**1945**	D945NDB	**1965**	D965PJA	**1980**	E980SJA
1906	D906NDB	**1929**	D929NDB	**1946**	D946NDB	**1966**	D966PJA	**1983**	E983SJA
1907	D907NDB	**1930**	D930NDB	**1947**	D947NDB	**1967**	D967PJA	**1984**	E984SJA
1908	D908NDB	**1931**	D931NDB	**1948**	D948NDB	**1969**	D968PJA	**1985**	E985SJA
1909	D909NDB	**1932**	D932NDB	**1949**	D949NDB	**1970**	D970PJA	**1986**	E986SJA
1910	D910NDB	**1933**	D933NDB	**1950**	D950NDB	**1972**	D972PJA	**1987**	E987SJA
1914	D914NDB	**1934**	D934NDB	**1952**	D952NDB	**1973**	D973PJA	**1989**	E989SJA
1916	D916NDB	**1935**	D935NDB	**1954**	D954NDB	**1975**	D975PJA	**1991**	E991SJA
1918	D918NDB	**1936**	D936NDB	**1956**	D956PJA	**1976**	D976PJA	**1994**	E994SJA
1919	D919NDB	**1940**	D940NDB	**1958**	D958PJA	**1977**	D977PJA	**1995**	E995SJA
1920	D920NDB	**1941**	D941NDB	**1960**	D960PJA	**1978**	D978PJA	**1996**	E996SJA
1925	D925NDB	**1942**	D942NDB	**1961**	D961PJA	**1979**	D979PJA	**1997**	E997SJA
1927	D927NDB	**1943**	D943NDB						

3011-3015		Leyland Olympian ONTL11/1R		Northern Counties		H43/30F	1983		
3011	A576HDB	**3012**	A577HDB	**3013**	A578HDB	**3014**	A579HDB	**3015**	A580HDB

3037-3238		Leyland Olympian ONLXB/1R		Northern Counties		H43/30F	1984-86		
3037	B37PJA	**3073**	B73PJA	**3108**	B108SJA	**3152**	B152XNA	**3202**	C202CBU
3038	B38PJA	**3075**	B75PJA	**3109**	B109SJA	**3157**	C157YBA	**3203**	C203CBU
3040	B40PJA	**3076**	B76PJA	**3111**	B111SJA	**3159**	C159YBA	**3204**	C204CBU
3041	B41PJA	**3078**	B78PJA	**3112**	B112SJA	**3160**	C160YBA	**3206**	C206CBU
3042	B42PJA	**3079**	B79PJA	**3113**	B113SJA	**3161**	C161YBA	**3209**	C209CBU
3043	B43PJA	**3081**	B81PJA	**3115**	B115SJA	**3162**	C162YBA	**3211**	C211CBU
3044	B44PJA	**3083**	B83PJA	**3116**	B116TVU	**3163**	C163YBA	**3217**	C217CBU
3045	B45PJA	**3085**	B85PJA	**3120**	B120TVU	**3168**	C168YBA	**3218**	C218CBU
3046	B46PJA	**3090**	B90SJA	**3123**	B123TVU	**3171**	C171YBA	**3219**	C219CBU
3047	B47PJA	**3092**	B92SJA	**3127**	B127WNB	**3177**	C177YBA	**3220**	C220CBU
3048	B48PJA	**3093**	B93SJA	**3128**	B128WNB	**3180**	C180YBA	**3222**	C222CBU
3050	B350PJA	**3096**	B96SJA	**3129**	B129WNB	**3182**	C182YBA	**3223**	C223CBU
3051	B351PJA	**3097**	B97SJA	**3130**	B130WNB	**3183**	C183YBA	**3225**	C225CBU
3052	B52PJA	**3098**	B98SJA	**3131**	B131WNB	**3186**	C186YBA	**3227**	C227ENE
3054	B54PJA	**3099**	B99SJA	**3134**	B134WNB	**3187**	C187YBA	**3228**	C228ENE
3059	B59PJA	**3100**	B100SJA	**3136**	B136WNB	**3188**	C188YBA	**3229**	C229ENE
3061	B61PJA	**3101**	B101SJA	**3140**	B140WNB	**3189**	C189YBA	**3231**	C231ENE
3062	B62PJA	**3102**	B102SJA	**3141**	B141WNB	**3190**	C190YBA	**3232**	C232ENE
3063	B63PJA	**3103**	B103SJA	**3142**	B142WNB	**3192**	C192YBA	**3233**	C233ENE
3064	B64PJA	**3104**	B104SJA	**3144**	B144WNB	**3194**	C194YBA	**3235**	C235ENE
3066	B66PJA	**3105**	B105SJA	**3148**	B148XNA	**3200**	C200YBA	**3237**	C237EVU
3068	B68PJA	**3106**	B106SJA	**3151**	B151XNA	**3201**	C201CBU	**3238**	C238EVU
3071	B71PJA	**3107**	B107SJA						

3239-3276		Leyland Olympian ONLXB/1R		Northern Counties		DPH43/26F	1986-87		
3239	C239EVU	**3246**	C246FRJ	**3253**	C253FRJ	**3262**	D262JVR	**3270**	D270JVR
3240	C240EVU	**3247**	C247FRJ	**3254**	C254FRJ	**3263**	D263JVR	**3271**	D271JVR
3241	C241EVU	**3248**	C248FRJ	**3256**	D256JVR	**3264**	D264JVR	**3273**	D273JVR
3242	C242EVU	**3249**	C249FRJ	**3257**	D257JVR	**3265**	D265JVR	**3274**	D274JVR
3243	C243EVU	**3250**	C250FRJ	**3258**	D258JVR	**3266**	D266JVR	**3275**	D275JVR
3244	C244EVU	**3251**	C251FRJ	**3259**	D259JVR	**3267**	D267JVR	**3276**	D276JVR
3245	C245EVU	**3252**	C252FRJ	**3261**	D261JVR				

3278-3305		Leyland Olympian ONLXB/1RZ		Northern Counties		H43/30F	1988-89		
3278	F278DRJ	**3281**	F281DRJ	**3287**	F287DRJ	**3292**	F292DRJ	**3302**	F302DRJ
3279	F279DRJ	**3284**	F284DRJ	**3288**	F288DRJ	**3293**	F293DRJ	**3303**	F303DRJ
3280	F280DRJ	**3286**	F286DRJ	**3290**	F290DRJ	**3299**	F299DRJ	**3305**	F305DRJ

Representing GMNs allocation of Leyland Olympians is 3223, C223CBU, belonging to Bolton depot, it is seen on layover in Wigan bus station.
Mike Fowler

4003-4018

Leyland Fleetline FE30AGR | Northern Counties | H43/32F | 1978 | Ex Yorkshire Rider, 1994

4003	XBU3S	**4007**	XBU7S	**4013**	XBU13S	**4016**	XBU16S	**4018**	XBU18S
4004	XBU4S	**4012**	XBU12S						

4026-4096

Leyland Fleetline FE30AGR | Northern Counties | H43/32F | 1978-79 | 4026 ex Yorkshire Rider, 1994

4026	ANA26T	**4056**	BVR56T	**4064**	BVR64T	**4078**	BVR78T	**4090**	BVR90T
4039	ANA39T	**4057**	BVR57T	**4072**	BVR72T	**4079**	BVR79T	**4091**	BVR91T
4046	ANA46T	**4058**	BVR58T	**4073**	BVR73T	**4081**	BVR81T	**4093**	BVR93T
4050	ANA50T	**4060**	BVR60T	**4074**	BVR74T	**4084**	BVR84T	**4095**	BVR95T
4051	BVR51T	**4062**	BVR62T	**4075**	BVR75T	**4086**	BVR86T	**4096**	BVR96T
4054	BVR54T	**4063**	BVR63T	**4077**	BVR77T	**4088**	BVR88T		

4102-4150

Leyland Fleetline FE30AGR | Northern Counties | H43/32F | 1979-80

4102	HDB102V	**4113**	HDB113V	**4122**	HDB122V	**4131**	KDB131V	**4142**	GNF17V
4104	HDB104V	**4114**	HDB114V	**4123**	HDB123V	**4132**	KDB132V	**4143**	MNC486W
4105	HDB105V	**4115**	HDB115V	**4125**	HDB125V	**4133**	KDB133V	**4146**	MNC489W
4106	HDB106V	**4117**	HDB117V	**4126**	HDB126V	**4136**	KDB136V	**4147**	MNC480W
4108	HDB108V	**4118**	HDB118V	**4127**	KDB127V	**4139**	KDB139V	**4148**	MNC491W
4109	HDB109V	**4119**	HDB119V	**4128**	KDB128V	**4140**	KDB140V	**4149**	MNC492W
4110	HDB110V	**4120**	HDB120V	**4129**	KDB129V	**4141**	GNF16V	**4150**	MNC493W
4111	HDB111V	**4121**	HDB121V	**4130**	KDB130V				

4151-4199

Leyland Atlantean AN68A/1R | Northern Counties | H43/32F* | 1978-79 | *4172 is O43/32F

4151	VBA151S	**4160**	VBA160S	**4172**	VBA172S	**4183**	VBA183S	**4194**	VBA194S
4154	VBA154S	**4163**	VBA163S	**4177**	VBA177S	**4184**	VBA184S	**4196**	VBA196S
4155	VBA155S	**4165**	VBA165S	**4179**	VBA179S	**4189**	VBA189S	**4197**	VBA197S
4158	VBA158S	**4167**	VBA167S	**4180**	VBA180S	**4191**	VBA191S	**4198**	VBA198S
4159	VBA159S	**4171**	VBA171S	**4181**	VBA181S	**4192**	VBA192S	**4199**	VBA199S

4201-4300

Leyland Atlantean AN68A/1R | Northern Counties | H43/32F | 1979-80

4201	XRJ201S	**4226**	ANA226T	**4255**	FVR255V	**4269**	FVR269V	**4285**	FVR285V
4202	XRJ202S	**4228**	ANA228T	**4256**	FVR256V	**4271**	FVR271V	**4286**	FVR286V
4205	XRJ205S	**4229**	ANA229T	**4257**	FVR257V	**4272**	FVR272V	**4287**	FVR287V
4206	XRJ206S	**4232**	ANA232T	**4258**	FVR258V	**4273**	FVR273V	**4288**	FVR288V
4207	ANA207T	**4233**	ANA233T	**4260**	FVR260V	**4274**	FVR274V	**4289**	FVR289V
4208	ANA208T	**4238**	ANA238T	**4261**	FVR261V	**4275**	FVR275V	**4292**	FVR292V
4209	ANA209T	**4239**	ANA239T	**4262**	FVR262V	**4276**	FVR276V	**4293**	FVR293V
4213	ANA213T	**4240**	FVR240V	**4263**	FVR263V	**4277**	FVR277V	**4296**	FVR296V
4216	ANA216T	**4243**	FVR243V	**4265**	FVR265V	**4278**	FVR278V	**4297**	FVR297V
4219	ANA219T	**4244**	FVR244V	**4266**	FVR266V	**4279**	FVR279V	**4298**	FVR298V
4221	ANA221T	**4251**	FVR251V	**4267**	FVR267V	**4282**	FVR282V	**4299**	FVR299V
4223	ANA223T	**4253**	FVR253V	**4268**	FVR268V	**4284**	FVR284V	**4300**	FVR300V
4224	ANA224T	**4254**	FVR254V						

4303-4350

Leyland Atlantean AN68A/1R | Northern Counties | H43/32F | 1980-81

4303	KDB303V	**4315**	MNC515W	**4326**	MNC526W	**4336**	MNC536W	**4344**	MNC544W
4305	MNC505W	**4316**	MNC516W	**4330**	MNC530W	**4337**	MNC537W	**4345**	MNC545W
4306	MNC506W	**4318**	MNC518W	**4331**	MNC531W	**4338**	MNC538W	**4346**	MNC546W
4308	MNC508W	**4320**	MNC520W	**4333**	MNC533W	**4340**	MNC540W	**4347**	MNC547W
4310	MNC510W	**4322**	MNC522W	**4334**	MNC534W	**4341**	MNC541W	**4349**	MNC549W
4311	MNC511W	**4323**	MNC523W	**4335**	MNC535W	**4342**	MNC542W	**4350**	MNC550W
4313	MNC513W	**4324**	MNC524W						

4351-4400

Leyland Atlantean AN68A/1R | Northern Counties | H43/32F | 1981

4351	ORJ351W	**4360**	ORJ360W	**4371**	ORJ371W	**4383**	ORJ383W	**4390**	ORJ390W
4353	ORJ353W	**4361**	ORJ361W	**4372**	ORJ372W	**4384**	ORJ384W	**4391**	ORJ391W
4354	ORJ354W	**4363**	ORJ363W	**4373**	ORJ373W	**4385**	ORJ385W	**4392**	ORJ392W
4355	ORJ355W	**4364**	ORJ364W	**4374**	ORJ374W	**4386**	ORJ386W	**4394**	ORJ394W
4356	ORJ356W	**4366**	ORJ366W	**4378**	ORJ378W	**4387**	ORJ387W	**4398**	ORJ398W
4357	ORJ357W	**4367**	ORJ367W	**4379**	ORJ379W	**4388**	ORJ388W	**4399**	ORJ399W
4358	ORJ358W	**4369**	ORJ369W	**4382**	ORJ382W	**4389**	ORJ389W	**4400**	ORJ400W
4359	ORJ359W	**4370**	ORJ370W						

As the book was being prepared, the first repaints into the new GMN scheme were entering service. Working serive 26A from the new depot in Liverpool is 4018, XBU18S in the new scheme.
Andrew Jarosz

4401-4525

Leyland Atlantean AN68A/1R* Northern Counties H43/32F 1981-82
*4448/56/60-81/5/93/4/502-25 are AN68B/1R

4401	MRJ401W	**4426**	SND426X	**4449**	SND449X	**4478**	SND478X	**4503**	SND503X
4403	MRJ403W	**4427**	SND427X	**4456**	SND456X	**4481**	SND481X	**4504**	SND504X
4405	MRJ405W	**4428**	SND428X	**4457**	SND457X	**4482**	SND482X	**4507**	SND507X
4407	MRJ407W	**4431**	SND431X	**4458**	SND458X	**4485**	SND485X	**4508**	SND508X
4408	MRJ408W	**4433**	SND433X	**4459**	SND459X	**4486**	SND486X	**4509**	SND509X
4410	MRJ410W	**4434**	SND434X	**4460**	SND460X	**4490**	SND490X	**4510**	SND510X
4411	MRJ411W	**4436**	SND436X	**4461**	SND461X	**4491**	SND491X	**4511**	SND511X
4413	SND413X	**4438**	SND438X	**4463**	SND463X	**4492**	SND492X	**4515**	SND515X
4414	SND414X	**4439**	SND439X	**4464**	SND464X	**4493**	SND493X	**4516**	SND516X
4415	SND415X	**4441**	SND441X	**4466**	SND466X	**4494**	SND494X	**4517**	SND517X
4416	SND416X	**4442**	SND442X	**4469**	SND469X	**4497**	SND497X	**4520**	SND520X
4417	SND417X	**4443**	SND443X	**4470**	SND470X	**4498**	SND498X	**4522**	SND522X
4419	SND419X	**4444**	SND444X	**4471**	SND471X	**4499**	SND499X	**4523**	SND523X
4420	SND420X	**4445**	SND445X	**4474**	SND474X	**4500**	SND500X	**4524**	SND524X
4423	SND423X	**4446**	SND446X	**4475**	SND475X	**4502**	SND502X	**4525**	SND525X
4424	SND424X	**4447**	SND447X						

4528-4599

Leyland Atlantean AN68D/1R Northern Counties H43/32F 1982

4528	SND528X	**4542**	ANA542Y	**4557**	ANA557Y	**4571**	ANA571Y	**4584**	ANA584Y
4529	SND529X	**4544**	ANA544Y	**4558**	ANA558Y	**4573**	ANA573Y	**4587**	ANA587Y
4531	ANA531Y	**4547**	ANA547Y	**4560**	ANA560Y	**4574**	ANA574Y	**4588**	ANA588Y
4532	ANA532Y	**4548**	ANA548Y	**4561**	ANA561Y	**4575**	ANA575Y	**4590**	ANA590Y
4534	ANA534Y	**4549**	ANA549Y	**4562**	ANA562Y	**4576**	ANA577Y	**4591**	ANA591Y
4535	ANA535Y	**4551**	ANA551Y	**4563**	ANA563Y	**4578**	ANA578Y	**4594**	ANA594Y
4536	ANA536Y	**4554**	ANA554Y	**4566**	ANA566Y	**4580**	ANA580Y	**4595**	ANA595Y
4539	ANA539Y	**4555**	ANA555Y	**4567**	ANA567Y	**4581**	ANA581Y	**4598**	ANA598Y
4540	ANA540Y	**4556**	ANA556Y	**4570**	ANA570Y	**4583**	ANA583Y	**4599**	ANA599Y
4541	ANA541Y								

4602-4697		Leyland Atlantean AN68D/1R		Northern Counties		H43/32F	1982-84		
4602	ANA602Y	**4619**	ANA619Y	**4638**	ANA638Y	**4655**	ANA655Y	**4676**	A676HNB
4603	ANA603Y	**4621**	ANA621Y	**4640**	ANA640Y	**4656**	A656HNB	**4677**	A677HNB
4606	ANA606Y	**4622**	ANA622Y	**4641**	ANA641Y	**4658**	A658HNB	**4681**	A681HNB
4607	ANA607Y	**4623**	ANA623Y	**4642**	ANA642Y	**4659**	A659HNB	**4682**	A682HNB
4610	ANA610Y	**4626**	ANA626Y	**4643**	ANA643Y	**4662**	A662HNB	**4685**	A685HNB
4611	ANA611Y	**4628**	ANA628Y	**4645**	ANA645Y	**4663**	A663HNB	**4686**	A686HNB
4614	ANA614Y	**4629**	ANA629Y	**4648**	ANA648Y	**4666**	A666HNB	**4689**	A689HNB
4615	ANA615Y	**4633**	ANA633Y	**4649**	ANA649Y	**4667**	A667HNB	**4691**	A691HNB
4616	ANA616Y	**4634**	ANA634Y	**4650**	ANA650Y	**4670**	A670HNB	**4692**	A692HNB
4617	ANA617Y	**4635**	ANA635Y	**4652**	ANA652Y	**4672**	A672HNB	**4697**	A697HNB
4618	ANA618Y	**4636**	ANA636Y	**4654**	ANA654Y	**4673**	A673HNB		

4701-4765		Leyland Atlantean AN68D/1R		Northern Counties		H43/32F	1984		
4701	A701LNC	**4717**	A717LNC	**4728**	A728LNC	**4738**	A738NNA	**4755**	A755NNA
4703	A703LNC	**4718**	A718LNC	**4729**	A729LNC	**4739**	A739NNA	**4756**	A756NNA
4704	A704LNC	**4720**	A720LNC	**4731**	A731LNC	**4740**	A740NNA	**4758**	A758NNA
4707	A707LNC	**4721**	A721LNC	**4732**	A732LNC	**4742**	A742NNA	**4760**	A760NNA
4712	A712LNC	**4723**	A723LNC	**4733**	A733NNA	**4746**	A746NNA	**4763**	A763NNA
4713	A713LNC	**4724**	A724LNC	**4736**	A736NNA	**4753**	A753NNA	**4765**	A765NNA
4716	A716LNC	**4727**	A727LNC	**4737**	A737NNA				

4914	PTD642S	Leyland Fleetline FE30AGR	Northern Counties	H43/32F	1977	Ex Yorkshire Rider, 1994
4920	PTD648S	Leyland Fleetline FE30AGR	Northern Counties	H43/32F	1977	Ex Yorkshire Rider, 1994
4922	PTD650S	Leyland Fleetline FE30AGR	Northern Counties	H43/32F	1977	Ex Yorkshire Rider, 1994
4926	PTD654S	Leyland Fleetline FE30AGR	Northern Counties	H43/32F	1977	Ex Yorkshire Rider, 1994

4942-4990		Leyland Fleetline FE30AGR		Northern Counties		H43/32F	1978-80		
4942	TWH701T	**4952**	YTE589V	**4964**	BCB617V	**4973**	DWH689W	**4982**	DWH698W
4943	TWH702T	**4953**	YTE590V	**4965**	BCB618V	**4974**	DWH690W	**4985**	DWH701W
4944	TWH703T	**4955**	YTE592V	**4967**	DWH683W	**4975**	DWH691W	**4986**	DWH702W
4945	TWH704T	**4956**	YTE593V	**4969**	DWH685W	**4976**	DWH692W	**4987**	DWH703W
4946	WWH94T	**4961**	BCB614V	**4970**	DWH686W	**4977**	DWH693W	**4988**	DWH704W
4947	YTE584V	**4962**	BCB615V	**4971**	DWH687W	**4979**	DWH695W	**4989**	DWH705W
4948	YTE585V	**4963**	BCB616V	**4972**	DWH688W	**4981**	DWH697W	**4990**	DWH706W
4949	YTE586V								

In additon to Leyland Atlanteans, four Fleetlines have been re-purchased from Yorkshire Rider, and repainted into fleet livery. One of these is 4914, PTD642S, new to Lancashire United, it is seen in service with GMN at Wigan.
Andrew Jarosz

5011-5015		MCW Metrobus DR102/10		MCW		H43/30F	1980		
5011	GBU11V	**5012**	GBU12V	**5013**	GBU13V	**5014**	GBU14V	**5015**	GBU15V

5031-5105		MCW Metrobus DR102/21		MCW		H43/30F	1981		
5031	MRJ31W	**5057**	MRJ57W	**5064**	MRJ64W	**5085**	ORJ85W	**5097**	ORJ97W
5032	MRJ32W	**5058**	MRJ58W	**5068**	MRJ68W	**5086**	ORJ86W	**5099**	ORJ99W
5033	MRJ33W	**5059**	MRJ59W	**5069**	MRJ69W	**5087**	ORJ87W	**5101**	ORJ101W
5034	MRJ34W	**5060**	MRJ60W	**5070**	MRJ70W	**5088**	ORJ88W	**5102**	SND102X
5035	MRJ35W	**5061**	MRJ61W	**5082**	ORJ82W	**5089**	ORJ89W	**5103**	SND103X
5039	MRJ39W	**5062**	MRJ62W	**5083**	ORJ83W	**5090**	ORJ90W	**5104**	SND104X
5050	MRJ50W	**5063**	MRJ63W	**5084**	ORJ84W	**5096**	ORJ96W	**5105**	SND105X
5056	MRJ56W								

5112-5189		MCW Metrobus DR102/23		MCW		H43/30F	1981-83		
5112	SND112X	**5131**	SND131X	**5147**	SND147X	**5168**	ANA168Y	**5181**	ANA181Y
5113	SND113X	**5133**	SND133X	**5148**	SND148X	**5169**	ANA169Y	**5182**	ANA182Y
5114	SND114X	**5134**	SND134X	**5149**	SND149X	**5171**	ANA171Y	**5183**	ANA183Y
5115	SND115X	**5135**	SND135X	**5150**	SND150X	**5172**	ANA172Y	**5184**	ANA184Y
5122	SND122X	**5136**	SND136X	**5151**	ANA151Y	**5174**	ANA174Y	**5185**	ANA185Y
5126	SND126X	**5137**	SND137X	**5152**	ANA152Y	**5175**	ANA175Y	**5186**	ANA186Y
5127	SND127X	**5138**	SND138X	**5156**	ANA156Y	**5176**	ANA176Y	**5187**	ANA187Y
5128	SND128X	**5139**	SND139X	**5166**	ANA166Y	**5177**	ANA177Y	**5188**	ANA188Y
5129	SND129X	**5140**	SND140X	**5167**	ANA167Y	**5178**	ANA178Y	**5189**	ANA189Y
5130	SND130X	**5146**	SND146X						

5201-5210		MCW Metrobus DR132/8		Northern Counties		DPH43/29F	1986		
5201	C201FVU	**5203**	C203FVU	**5205**	C205FVU	**5207**	C207FVU	**5209**	C209FVU
5202	C202FVU	**5204**	C204FVU	**5206**	C206FVU	**5208**	C208FVU	**5210**	C210FVU

5301-5320		MCW Metrobus DR102/51		Northern Counties		DPH43/29F	1986-87		
5301	D301JVR	**5305**	D305JVR	**5309**	D309JVR	**5313**	D313LNB	**5317**	D317LNB
5302	D302JVR	**5306**	D306JVR	**5310**	D310JVR	**5314**	D314LNB	**5318**	D318LNB
5303	D303JVR	**5307**	D307JVR	**5311**	D311LNB	**5315**	D315LNB	**5319**	D319LNB
5304	D304JVR	**5308**	D308JVR	**5312**	D312LNB	**5316**	D316LNB	**5320**	D320LNB

7001-7008		Volvo Citybus B10M-50		Northern Counties		H45/32F	1991		
7001	H701GVM	**7003**	H703GVM	**7005**	H705GVM	**7007**	H707GVM	**7008**	H708GVM
7002	H702GVM	**7004**	H704GVM	**7006**	H706GVM				

7009	J709ONF	Volvo Citybus B10M-50	Northern Counties	H45/28FL	1992
7010	J710ONF	Volvo Citybus B10M-50	Northern Counties	H45/28FL	1992

The last pair of Volvo Citybus double-decks, 7009 & 7010, were fitted with lifts capable of allowing the wheelchair bound to travel on the buses. The first of these, 7009, J709ONF, shows little sign of the lift in this picture taken at Fleetwood Tram Sunday. Also visible is the lettering for its normal route, the 409 from Shaw, through Oldham to Stalybridge.
Andrew Jarosz

The only GMN depots with Metrobuses are Queens Road and Oldham. While eighty-three of the MCW-bodied variety are with GMN, all thirty Northern Counties examples are operated. All the latter are in the pink Express livery and have high-back seating. Left is 5171, ANA171Y, operating service 17 from Rochdale to Manchester, while below is 5317, D317LNB, one of the latest of the type seen passing Oldham Technical College on service 59.
Mike Fowler/
Roy Marshall

Bury depot's 7767, UNA767S is seen in Norden working service R1, normally operated by minibuses. One of the 1977 delivery of Leyland Atlanteans, has recently been joined by seven similar examples taken back into the GMN fleet from Yorkshire Rider to whom they were sold in 1988 by Greater Manchester. GMN have also taken delivery of their first new buses in the form of ten Volvo B6s with Northern Counties bodywork. *Ken Grime*

7077	WBN955L	Leyland Atlantean AN68/1R	Park Royal	O43/32F	1972	
7685	ONF685R	Leyland Atlantean AN68A/1R	Northern Counties	H43/32F	1976	Ex Yorkshire Rider, 1994
7691	ONF691R	Leyland Atlantean AN68A/1R	Northern Counties	H43/32F	1976	Ex Yorkshire Rider, 1994
7699	ONF699R	Leyland Atlantean AN68A/1R	Northern Counties	H43/32F	1977	

7706-7797 Leyland Atlantean AN68A/1R Northern Counties H43/32F 1977 7706-8/25/62/73/97 ex Yorkshire Rider, 1994

7706	RJA719R	**7724**	RJA724R	**7767**	UNA767S	**7779**	UNA779S	**7785**	UNA785S
7707	RJA719R	**7725**	RJA756R	**7768**	UNA768S	**7780**	UNA780S	**7787**	UNA787S
7708	RJA719R	**7726**	RJA726R	**7769**	UNA769S	**7782**	UNA782S	**7791**	UNA791S
7719	RJA719R	**7760**	SRJ760R	**7771**	UNA771S	**7783**	UNA783S	**7793**	UNA793S
7721	RJA721R	**7762**	UNA762S	**7773**	UNA773S	**7784**	UNA784S	**7797**	UNA797S

7802-7902 Leyland Atlantean AN68A/1R Park Royal H43/32F 1977-78

7802	RJA802R	**7821**	UNA821S	**7833**	UNA833S	**7846**	UNA846S	**7867**	UNA867S
7803	RJA803R	**7822**	UNA822S	**7834**	UNA834S	**7847**	UNA847S	**7883**	WVM883S
7805	RJA805R	**7828**	UNA828S	**7835**	UNA835S	**7852**	UNA852S	**7886**	WVM886S
7806	RJA806R	**7829**	UNA829S	**7837**	UNA837S	**7854**	UNA854S	**7898**	WVM898S
7809	RJA809R	**7830**	UNA830S	**7839**	UNA839S	**7866**	UNA866S	**7902**	WVM902S

7904-7951 Leyland Atlantean AN68A/1R Park Royal H43/32F 1978-79

7904	ANC904T	**7921**	ANC921T	**7928**	ANC928T	**7934**	BNC934T	**7949**	BNC949T
7911	ANC911T	**7922**	ANC922T	**7933**	BNC933T	**7937**	BNC937T	**7951**	BNC951T
7914	ANC914T								

Livery: Orange, white, grey and black; pink, orange, red, brown and white (Express) 3239-3277; 5201-5210; 5301-5320

GMS BUSES

Greater Manchester Buses South Ltd, Charterplan House, 151 Charles Street, Stockport, Greater Manchester, SK1 3JU

Depots: Birkenhead; Bennett Street, Manchester; Charles Street, Stockport; Hyde Road, Manchester; York Street, Glossop; Princess Road, Manchester and Daw Bank, Stockport.

1	OJI9451	Leyland Tiger TRCTL11/3R	Plaxton Paramount 3200 II	C53F	1985	
2	OJI9452	Leyland Tiger TRCTL11/3R	Plaxton Paramount 3200 II	C53F	1985	
3	NIW1673	Leyland Tiger TRCTL11/3RH	Plaxton Paramount 3200 II	C53F	1985	
4	TPX884	Leyland Tiger TRCTL11/3RH	Plaxton Paramount 3500 II	C49FT	1985	
5	OJI9455	Volvo B10M-61	Plaxton Paramount 3200 II	C53F	1985	
6	NIW1676	Volvo B10M-61	Plaxton Paramount 3200 II	C53F	1986	
7	IIL1047	Volvo B10M-61	Plaxton Paramount 3200 II	C53F	1986	
8	OIW1608	Leyland Tiger TRCTL11/3RH	Duple 320	C57F	1986	
9	NIW2399	Leyland Tiger TRCTL11/3RH	Duple 320	C57F	1986	
10	C310ENA	Leyland Tiger TRCTL11/3RH	Duple 320	C57F	1986	
11	SIB1361	Leyland Tiger TRCTL11/3RH	Duple 320	C57F	1986	
12	SIB1832	Leyland Tiger TRCTL11/3RH	Duple 320	C57F	1986	
14	SIB2014	Volvo B10M-61	Plaxton Paramount 3500 II	C49F	1986	Ex Excelsior, 1989
15	PXI7915	Volvo B10M-61	Plaxton Paramount 3500 III	C53F	1988	Ex Park's, Hamilton, 1989
16	PXI8916	Volvo B10M-61	Plaxton Paramount 3500 III	C53F	1988	Ex Park's, Hamilton, 1989
17	OIW1317	Leyland Tiger TRCTL11/3RH	Plaxton Paramount 3500 III	C49FT	1989	
18	OIW1318	Volvo B10M-61	Plaxton Paramount 3500 III	C53F	1989	Ex Park's, Hamilton, 1990
19	OIW1319	Volvo B10M-61	Plaxton Paramount 3500 III	C53F	1989	Ex Park's, Hamilton, 1990
20	H170DVL	Kässbohrer Setra S215HD	Kässbohrer Tornado	C49F	1990	
22	G181VBB	Kässbohrer Setra S215HD	Kässbohrer Tornado	C49F	1989	Ex Craiggs, Radcliffe, 1991
23	G715TTY	Kässbohrer Setra S215HD	Kässbohrer Tornado	C49F	1989	Ex Craiggs, Radcliffe, 1991
24	J73VTG	Kässbohrer Setra S215HD	Kässbohrer Tornado	C49F	1994	

The Charterplan coach unit is now part of GMS Buses and continues to operate out of Charles Street, the headquarters of the South operation. Included in the fleet are seven Kässbohrer Setra coaches, all with the Tornado bodywork design as shown by 20, H170DVL. *Andrew Jarosz*

Princess Road depot has become the home of the Dennis Dominators, all of which have Northern Counties bodywork. Photographed at the British Rail depot in the Longsight district of Manchester is 2040, H140GVM, though these vehicles are more normally to be found heading for Wythenshaw. *Cliff Beeton*

Opposite, top: **GMS Buses have made several changes since the separation of the fleets. Most notable is the introduction of twenty-three Leyland Nationals, including the first vehicle to leave the Leyland National plant (chassis 102), which had overtaken SELNEC's own vehicle (chassis 101) in the paintbooth! First of these vehicles is 251, RUF41R, seen on service 192 in Stockport.** *Andrew*

Opposite, bottom: **Repainted into Birkenhead & District Livery is 4135, KDB135V, photographed in Liscard Village shortly after GMB South commenced their operations on the Wirral.** *John Young*

25	L543YUS	Volvo B10M-60	Van Hool Alizée	C53F	1989	Ex Park's, Hamilton, 1994
26	G174RBD	Volvo B10M-60	Jonckheere Deauville P599	C53F	1989	Ex Redwing, Camberwell, 1994
30	515VTB	Kässbohrer Setra S215HD	Kässbohrer Tornado	C49F	1983	
32	OTK802	Kässbohrer Setra S215HD	Kässbohrer Tornado	C49F	1983	
33	OXK373	Kässbohrer Setra S215HD	Kässbohrer Tornado	C49F	1983	
64	583TD	Leyland Tiger TRCTL11/3R	Plaxton Paramount 3200	C53F	1983	Ex The Londoners, Nunhead, 1985
65	KGS493Y	Leyland Tiger TRCTL11/3R	Plaxton Paramount 3200	C53F	1983	Ex The Londoners, Nunhead, 1985
82	SND82X	Leyland Leopard PSU3B/4R	Duple Dominant IV (1981)	C51F	1975	
86	SND86X	Leyland Leopard PSU3B/4R	Duple Dominant IV (1982)	C51F	1975	

100	476CEL	Leyland Leopard PSU3B/4R	Duple Dominant IV (1984)	C34FTL	1975	
160	KBU895P	Leyland National 10351/1R		B41F	1975	

190-240 Leyland National 11351A/1R B49F* 1977/79 *190 is DP42F

190	ABA15T	**194**	ABA19T	**205**	ABA30T	**232**	WBN466T	**240**	WBN474T
192	ABA17T	**199**	ABA24T						

251	RUF41R	Leyland National 11351A/2R		B49F*	1977	Ex Hogg, Glasgow, 1994
252	DBY718M	Leyland National 1151/1R/0401		B49F	1973	Ex Munro, Uddington, 1994
253	DBY717M	Leyland National 1151/1R/0402		B49F	1973	Ex R & I, Milton Koynes, 1994
254	KRE295P	Leyland National 11351/1R		B52F	1976	Ex Rose Hill, Marple, 1994
255	EUM900T	Leyland National 11351A/1R		B49F	1979	Ex Amberley, Pudsey, 1994
256	FNS161T	Leyland National 11351A/1R		B49F	1978	Ex Amberley, Pudsey, 1994
257	CFM352S	Leyland National 11351A/1R		B49F	1978	Ex Amberley, Pudsey, 1994
258	LPR937P	Leyland National 11351/1R		B48F	1975	Ex Amberley, Pudsey, 1994
259	HMA657N	Leyland National 1151/1R/SC		DP49F	1975	Ex Amberley, Pudsey, 1994
260	JHU861L	Leyland National 1151/1R/0403		B44D	1973	Ex Amberley, Pudsey, 1994
261	KDW347P	Leyland National 11351/1R/SC		DP48F	1975	Ex Golden Coaches, Llantwit Major, 1994
262	KDW351P	Leyland National 11351/1R/SC		DP48F	1975	Ex Red & White, 1994
263	HFM179N	Leyland National 11351/1R/SC		DP48F	1974	Ex ??, 1994
264	KDW342P	Leyland National 11351/1R		B49F	1975	Ex Gatwick Handling, 1994
265	WFM801K	Leyland National 1151/2R/0403		B44D	1972	Ex Gatwick Handling, 1994
266	WFM803L	Leyland National 1151/2R/0403		B44D	1972	Ex Gatwick Handling, 1994
267	WFM806L	Leyland National 1151/2R/0403		B44D	1972	Ex Gatwick Handling, 1994
268	WFM811L	Leyland National 1151/2R/0403		B44D	1972	Ex Gatwick Handling, 1994
269	WFM817L	Leyland National 1151/2R/0403		B44D	1972	Ex Gatwick Handling, 1994
270	WFM822L	Leyland National 1151/2R/0403		B44D	1973	Ex Gatwick Handling, 1994
271	XRB416L	Leyland National 1151/2R/0403		B44D	1973	Ex Gatwick Handling, 1994
272	JHU868L	Leyland National 1151/2R/0403		B44D	1973	Ex Gatwick Handling, 1994
273	GFJ663N	Leyland National 11351/2R		B46D	1974	Ex Gatwick Handling, 1994

192 MANCHESTER
GM buses SOUTH
251
GM buses
LEYLAND
RUF 41R

SAUNA
AEROBICS
BIRKENHEAD & DISTRICT
3B
ARROWE PARK HOSP
4135
BIRKENHEAD & DISTRICT
KDB 135V
MORETON CROSS
AMOUS

1451	NJA568W	Leyland Olympian B45/TL11/1R	Northern Counties	H43/30F	1980
1461	FWH461Y	Scania BR112DH	Northern Counties	H43/32F	1983
1462	FWH462Y	Scania BR112DH	Northern Counties	H43/32F	1983

1463-1467 Scania N113DRB — Northern Counties — H47/28F — 1991

1463	H463GVM	**1464**	H464GVM	**1465**	H465GVM	**1466**	H466GVM	**1467**	H467GVM

1471	A471HNC	Dennis Falcon V	Northern Counties	H47/37F	1984
1472	A472HNC	Dennis Falcon V	Northern Counties	H47/37F	1984
1473	A473HNC	Dennis Falcon V	Northern Counties	H47/37F	1984
1504	D504MJA	Iveco Daily 49.10	Robin Hood City Nippy	B19F	1987

1601-1680 MCW Metrorider MF151/3 — MCW — B23F* — 1987 — *1639 is DP19F

1601	D601MDB	**1617**	D617MDB	**1635**	D635MDB	**1651**	D651NNE	**1666**	D666NNE
1602	D602MDB	**1618**	D618MDB	**1636**	D636MDB	**1652**	D652NNE	**1667**	D667NNE
1603	D603MDB	**1619**	D619MDB	**1638**	D638MDB	**1653**	D653NNE	**1668**	D668NNE
1604	D604MDB	**1620**	D620MDB	**1639**	D639MDB	**1654**	D654NNE	**1670**	D670NNE
1605	D605MDB	**1621**	D621MDB	**1640**	D640MDB	**1655**	D655NNE	**1671**	D671NNE
1606	D606MDB	**1623**	D623MDB	**1641**	D641MDB	**1656**	D656NNE	**1672**	D672NNE
1607	D607MDB	**1624**	D624MDB	**1642**	D642MDB	**1657**	D657NNE	**1673**	D673NNE
1608	D608MDB	**1626**	D626MDB	**1643**	D643MDB	**1658**	D658NNE	**1674**	D674NNE
1609	D609MDB	**1627**	D627MDB	**1644**	D644MDB	**1659**	D659NNE	**1675**	D675NNE
1610	D610MDB	**1628**	D628MDB	**1645**	D645MDB	**1660**	D660NNE	**1676**	D676NNE
1611	D611MDB	**1629**	D629MDB	**1646**	D646MDB	**1661**	D661NNE	**1677**	D677NNE
1612	D612MDB	**1630**	D630MDB	**1647**	D647MDB	**1662**	D662NNE	**1678**	D678NNE
1613	D613MDB	**1632**	D632MDB	**1648**	D648MDB	**1663**	D663NNE	**1679**	D679NNE
1614	D614MDB	**1633**	D633MDB	**1649**	D649MDB	**1664**	D664NNE	**1680**	D680NNE
1615	D615MDB	**1634**	D634MDB	**1650**	D650MDB	**1665**	D665NNE		

1721-1728 Renault-Dodge S75 — Northern Counties — B17FL — 1990-91

1721	H721CNC	**1724**	H724CNC	**1726**	H726CNC	**1727**	H727FNC	**1728**	H728FNC

1751-1770 Dennis Domino SDA1201 — Northern Counties — B24F — 1985-86

1751	C751YBA	**1755**	C755YBA	**1760**	C760YBA	**1764**	C764YBA	**1768**	C768YBA
1752	C752YBA	**1756**	C756YBA	**1761**	C761YBA	**1765**	C765YBA	**1769**	C769YBA
1753	C753YBA	**1757**	C757YBA	**1762**	C762YBA	**1766**	C766YBA	**1770**	C770YBA
1754	C754YBA	**1758**	C758YBA	**1763**	C763YBA	**1767**	C767YBA		

1772-1898 Renault-Dodge S56 — Northern Counties — B20F — 1986-87

1772	D772RBU	**1781**	D781RBU	**1791**	E791SJA	**1800**	E800SJA	**1891**	D891MDB
1773	D773RBU	**1782**	D783RBU	**1792**	E792SJA	**1843**	D843LND	**1892**	D892MDB
1774	D774RBU	**1787**	E787SJA	**1793**	E793SJA	**1861**	D861LND	**1896**	D896MDB
1775	D775RBU	**1788**	E788SJA	**1797**	E797SJA	**1865**	D865LND	**1898**	D898MDB
1776	D776RBU	**1789**	E789SJA	**1798**	E798SJA				

1917-2000 Renault-Dodge S56 — Northern Counties — B20F — 1987

1911	D911NDB	**1923**	D923NDB	**1953**	D953NDB	**1968**	D968PJA	**1992**	E992SJA
1912	D912NDB	**1926**	D926NDB	**1957**	D957PJA	**1971**	D971PJA	**1993**	E993SJA
1913	D913NDB	**1938**	D938NDB	**1959**	D959PJA	**1974**	D974PJA	**1998**	E998SJA
1917	D917NDB	**1939**	D939NDB	**1962**	D962PJA	**1981**	E981SJA	**1999**	E999SJA
1921	D921NDB	**1944**	D944NDB	**1963**	D963PJA	**1988**	E988SJA	**2000**	E200SVR
1922	D922NDB	**1951**	D951NDB	**1964**	D964PJA	**1990**	E990SJA		

2001-2030 Dennis Dominator DDA1003 — Northern Counties — H43/32F — 1985

2001	B901TVR	**2007**	B907TVR	**2013**	B913TVR	**2019**	B919TVR	**2025**	B25TVU
2002	B902TVR	**2008**	B908TVR	**2014**	B914TVR	**2020**	B920TVR	**2026**	B26TVU
2003	B903TVR	**2009**	B909TVR	**2015**	B915TVR	**2021**	B21TVU	**2027**	B27TVU
2004	B904TVR	**2010**	B910TVR	**2016**	B916TVR	**2022**	B22TVU	**2028**	B28TVU
2005	B905TVR	**2011**	B911TVR	**2017**	B917TVR	**2023**	B23TVU	**2029**	B29TVU
2006	B906TVR	**2012**	B912TVR	**2018**	B918TVR	**2024**	B24TVU	**2030**	B30TVU

2031-2040 Dennis Dominator DDA2033 — Northern Counties — H43/29F* — 1991 — *2032 is H42/29F

2031	H131GVM	**2033**	H133GVM	**2035**	H135GVM	**2037**	H137GVM	**2039**	H139GVM
2032	H132GVM	**2034**	H134GVM	**2036**	H136GVM	**2038**	H138GVM	**2040**	H140GVM

The Scania products in the GM Buses fleet were allocated to Hyde Road at the time of the division, and thus have become part of the GMS Buses fleet. Photographed outside the tiled Arndale Centre is 1462, FWH462Y, one of two BR112DH types that spent their early days at Atherton, the later five being type N113DRB. *Phillip Stephenson*

Vehicles have started to be painted in a new livery for GMS Buses which includes black window surounds. Seen here is 1653, D653NNE, one of the large batch delivered new in 1987.
Andrew Jarosz

3001-3010	Leyland Olympian ONTL11/1R	Northern Counties	H43/30F	1982

3001	ANA1Y	**3003**	ANA3Y	**3005**	ANA5Y	**3007**	ANA7Y	**3009**	ANA9Y
3002	ANA2Y	**3004**	ANA4Y	**3006**	ANA6Y	**3008**	ANA8Y	**3010**	ANA10Y

3016-3025	Leyland Olympian ONLXCT/1R	Northern Counties	H43/30F	1983-84

3016	A581HDB	**3018**	A583HDB	**3020**	A585HDB	**3022**	A22HNC	**3024**	A24HNC
3017	A582HDB	**3019**	A584HDB	**3021**	A21HNC	**3023**	A23HNC	**3025**	A25HNC

3026-3035	Leyland Olympian ONLXB/1R	Northern Counties	H43/30F	1984

3026	A26ORJ	**3028**	A28ORJ	**3030**	A30ORJ	**3032**	A32ORJ	**3034**	B34PJA
3027	A27ORJ	**3029**	A29ORJ	**3031**	A31ORJ	**3033**	A33ORJ	**3035**	B35PJA

3036-3238	Leyland Olympian ONLXB/1R	Northern Counties	H43/30F*	1984-86 *3139 is DPH43/15F 3198 and 3213 are DPH43/26F; 3214 is DPH43/18F

3036	B36PJA	**3084**	B84PJA	**3132**	B132WNB	**3165**	C165YBA	**3195**	C195YBA
3039	B39PJA	**3086**	B86SJA	**3133**	B133WNB	**3166**	C166YBA	**3198**	C198YBA
3049	B49PJA	**3087**	B87SJA	**3135**	B135WNB	**3167**	C167YBA	**3205**	C205CBU
3053	B53PJA	**3088**	B88SJA	**3137**	B137WNB	**3169**	C169YBA	**3207**	C207CBU
3055	B55PJA	**3089**	B89SJA	**3138**	B138WNB	**3170**	C170YBA	**3208**	C208CBU
3056	B56PJA	**3091**	B91SJA	**3139**	B139WNB	**3172**	C172YBA	**3210**	C210CBU
3057	B57PJA	**3094**	B94SJA	**3143**	B143WNB	**3173**	C173YBA	**3212**	C212CBU
3058	B58PJA	**3095**	B95SJA	**3145**	B145WNB	**3174**	C174YBA	**3213**	C213CBU
3060	B60PJA	**3110**	B110SJA	**3146**	B146XNA	**3175**	C175YBA	**3214**	C214CBU
3065	B65PJA	**3114**	B114SJA	**3147**	B147XNA	**3176**	C176YBA	**3215**	C215CBU
3067	B67PJA	**3117**	B117TVU	**3149**	B149XNA	**3178**	C178YBA	**3216**	C216CBU
3069	B69PJA	**3118**	B118TVU	**3150**	B150XNA	**3179**	C179YBA	**3221**	C221CBU
3070	B70PJA	**3119**	B119TVU	**3153**	B153XNA	**3181**	C181YBA	**3224**	C224CBU
3072	B72PJA	**3121**	B121TVU	**3154**	B154XNA	**3184**	C184YBA	**3226**	C226ENE
3074	B74PJA	**3122**	B122TVU	**3155**	B155XNA	**3185**	C185YBA	**3230**	C230ENE
3077	B77PJA	**3124**	B124TVU	**3156**	C156YBA	**3191**	C191YBA	**3234**	C234ENE
3080	B80PJA	**3125**	B125TVU	**3158**	C158YBA	**3193**	C193YBA	**3236**	C236EVU
3082	B82PJA	**3126**	B126WNB	**3164**	C164YBA				

3255-3277	Leyland Olympian ONLXB/1R	Northern Counties	DPH43/26F	1986-87

3255	C255FRJ	**3268**	D268JVR	**3269**	D269JVR	**3272**	D272JVR	**3277**	D277JVR
3260	D260JVR								

3282-3304	Leyland Olympian ONLXB/1RZ	Northern Counties	H43/30F	1988-89

3282	F282DRJ	**3289**	F289DRJ	**3295**	F295DRJ	**3298**	F298DRJ	**3301**	F301DRJ
3283	F283DRJ	**3291**	F291DRJ	**3296**	F296DRJ	**3300**	F300DRJ	**3304**	F304DRJ
3285	F285DRJ	**3294**	F294DRJ	**3297**	F297DRJ				

4002	XBU20S	Leyland Fleetline FE30AGR	Northern Counties	H43/32F	1978
4020	XBU20S	Leyland Fleetline FE30AGR	Northern Counties	H43/32F	1978

4021-4100	Leyland Fleetline FE30AGR	Northern Counties	H43/32F	1978-79

4021	ANA21T	**4035**	ANA35T	**4042**	ANA42T	**4061**	BVR61T	**4098**	BVR98T
4024	ANA24T	**4037**	ANA37T	**4043**	ANA43T	**4068**	BVR68T	**4100**	BVR100T

4101-4145	Leyland Fleetline FE30AGR	Northern Counties	H43/32F	1979-80

4101	HDB101V	**4107**	HDB107V	**4135**	KDB135V	**4138**	KDB138V	**4145**	MNC488W
4103	HDB103V	**4116**	HDB116V	**4137**	KDB137V	**4144**	MNC487W		

4152-4200	Leyland Atlantean AN68A/1R	Northern Counties	H43/32F*	1978 *4172 is O43/32F

4152	VBA152S	**4164**	VBA164S	**4170**	VBA170S	**4182**	VBA182S	**4190**	VBA190S
4153	VBA153S	**4166**	VBA166S	**4174**	VBA174S	**4186**	VBA186S	**4193**	VBA193S
4156	VBA156S	**4168**	VBA168S	**4175**	VBA175S	**4187**	VBA187S	**4195**	VBA195S
4157	VBA157S	**4169**	VBA169S	**4178**	VBA178S	**4188**	VBA188S	**4200**	VBA200S
4162	VBA162S								

Several of the Leyland Atlanteans and Olympians have recently been repainted, and their foglamps replaced by black discs. Typical is 3039, B39PJA, photographed in Stockport bus station while working service 371 to Altrincham.
Andrew Jarosz

The 1988 models of the Leyland Olympian carry the later version of the Northern Counties body with modified front similar to the early version of the Palatine. Since the closure of the Tameside depot, Hyde Road vehicles are more common in the district where 3301, F301DRJ is seen while working service 348 to Micklehurst.
Andrew Jarosz

Representing the 'Standard' Greater Manchester bus is 4665, A665HNB, photographed in Deansgate while working service 50 to East Didsbury. This service once connected Parrs Wood depot and the city. When that depot closed the service was transferred to Birchfields Road, and has since moved to Hyde Road.
Andrew Jarosz

4203-4295

Leyland Atlantean AN68A/1R Northern Counties H43/32F 1978-80

4203	XRJ203S	**4218**	ANA218T	**4234**	ANA234T	**4246**	FVR246V	**4280**	FVR280V
4204	XRJ204S	**4220**	ANA220T	**4235**	ANA235T	**4247**	FVR247V	**4281**	FVR281V
4210	ANA210T	**4222**	ANA222T	**4236**	ANA236T	**4249**	FVR249V	**4283**	FVR283V
4212	ANA212T	**4225**	ANA225T	**4237**	ANA237T	**4250**	FVR250V	**4290**	FVR290V
4214	ANA214T	**4227**	ANA227T	**4241**	ANA241T	**4252**	FVR252V	**4291**	FVR291V
4215	ANA215T	**4230**	ANA230T	**4242**	FVR242V	**4259**	FVR259V	**4294**	FVR294V
4217	ANA217T	**4231**	ANA231T	**4245**	FVR245V	**4264**	FVR264V	**4295**	FVR295V

4301-4350

Leyland Atlantean AN68A/1R Northern Counties H43/32F 1980-81

4301	KDB301V	**4309**	MNC509W	**4319**	MNC519W	**4328**	MNC528W	**4339**	MNC539W
4302	KDB302V	**4312**	MNC512W	**4321**	MNC521W	**4329**	MNC529W	**4343**	MNC543W
4304	MNC504W	**4314**	MNC514W	**4325**	MNC525W	**4332**	MNC523W	**4348**	MNC548W
4307	MNC507W	**4317**	MNC517W	**4327**	MNC527W				

4352-4396

Leyland Atlantean AN68A/1R Northern Counties H43/32F 1981

4352	ORJ352W	**4368**	ORJ368W	**4377**	ORJ377W	**4381**	ORJ381W	**4395**	ORJ395W
4362	ORJ362W	**4375**	ORJ375W	**4380**	ORJ380W	**4393**	ORJ393W	**4396**	ORJ396W
4365	ORJ365W	**4376**	ORJ376W						

4402-4521

Leyland Atlantean AN68A/1R* Northern Counties H43/32F 1981-82

*4425/49/55-72/6/7/80-4, 501-21 are AN68B/1R

4402	MRJ402W	**4430**	SND430X	**4454**	SND454X	**4479**	SND479X	**4501**	SND501X
4404	MRJ404W	**4432**	SND434X	**4455**	SND455X	**4480**	SND480X	**4505**	SND505X
4406	MRJ406W	**4435**	SND435X	**4462**	SND452X	**4483**	SND483X	**4506**	SND506X
4409	MRJ409W	**4437**	SND437X	**4465**	SND465X	**4484**	SND484X	**4512**	SND512X
4412	SND412X	**4440**	SND440X	**4468**	SND468X	**4487**	SND487X	**4513**	SND513X
4418	SND418X	**4449**	SND449X	**4472**	SND472X	**4488**	SND488X	**4514**	SND514X
4421	SND421X	**4450**	SND450X	**4473**	SND473X	**4489**	SND489X	**4518**	SND518X
4422	SND422X	**4451**	SND451X	**4476**	SND476X	**4495**	SND495X	**4519**	SND519X
4425	SND425X	**4452**	SND452X	**4477**	SND477X	**4496**	SND496X	**4521**	SND521X
4429	SND429X	**4453**	SND453X						

4526-4600

Leyland Atlantean AN68D/1R Northern Counties H43/32F 1982

4526	SND526X	**4543**	ANA543Y	**4559**	ANA559Y	**4579**	ANA579Y	**4592**	ANA592Y
4527	SND527X	**4545**	ANA544Y	**4564**	ANA564Y	**4582**	ANA582Y	**4593**	ANA593Y
4530	SND530X	**4546**	ANA546Y	**4568**	ANA568Y	**4585**	ANA585Y	**4596**	ANA596Y
4533	ANA533Y	**4550**	ANA550Y	**4569**	ANA569Y	**4586**	ANA586Y	**4597**	ANA597Y
4537	ANA537Y	**4552**	ANA552Y	**4572**	ANA572Y	**4589**	ANA589Y	**4600**	ANA600Y
4538	ANA538Y	**4553**	ANA553Y	**4577**	ANA577Y				

4601-4700

Leyland Atlantean AN68D/1R Northern Counties H43/32F 1983-84

4601	ANA601Y	**4627**	ANA627Y	**4651**	ANA651Y	**4671**	A671HNB	**4688**	A688HNB
4604	ANA604Y	**4630**	ANA630Y	**4653**	ANA653Y	**4674**	A674HNB	**4690**	A690HNB
4605	ANA605Y	**4631**	ANA631Y	**4657**	A657HNB	**4675**	A675HNB	**4693**	A693HNB
4608	ANA608Y	**4632**	ANA633Y	**4660**	A660HNB	**4678**	A678HNB	**4694**	A694HNB
4609	ANA609Y	**4637**	ANA637Y	**4661**	A661HNB	**4679**	A679HNB	**4695**	A695HNB
4612	ANA612Y	**4639**	ANA639Y	**4664**	A664HNB	**4680**	A680HNB	**4696**	A696HNB
4613	ANA613Y	**4644**	ANA644Y	**4665**	A665HNB	**4683**	A683HNB	**4698**	A698HNB
4620	ANA620Y	**4646**	ANA646Y	**4668**	A668HNB	**4684**	A684HNB	**4699**	A699HNB
4624	ANA624Y	**4647**	ANA647Y	**4669**	A669HNB	**4687**	A687HNB	**4700**	A700HNB
4625	ANA625Y								

4702-4764

Leyland Atlantean AN68D/1R Northern Counties H43/32F 1984

4702	A702LNC	**4714**	A714LNC	**4734**	A734NNA	**4747**	A747NNA	**4754**	A754NNA
4704	A704LNC	**4715**	A715LNC	**4735**	A735NNA	**4748**	A748NNA	**4757**	A757NNA
4705	A705LNC	**4719**	A719LNC	**4741**	A741NNA	**4749**	A749NNA	**4759**	A759NNA
4706	A706LNC	**4722**	A722LNC	**4743**	A743NNA	**4750**	A750NNA	**4761**	A761NNA
4708	A708LNC	**4725**	A725LNC	**4744**	A744NNA	**4751**	A751NNA	**4762**	A762NNA
4710	A710LNC	**4726**	A726LNC	**4745**	A745NNA	**4752**	A752NNA	**4764**	A764NNA
4711	A711LNC	**4730**	A730LNC						

The MCW Metrobus was allocated to selected depots, mostly to the south and eastern part of the county, resulting in some being based at all GMS Buses depots. Photographed in Hyde, while working the Manchester to Glossop service 236, is 5117, SND117X. *Andrew Jarosz*

5017-5030		MCW Metrobus DR102/10		MCW		H43/30F	1980		
5017	GBU17V	**5020**	GBU20V	**5023**	MNC497W	**5027**	GBU27V	**5029**	GBU29V
5018	MNC494W	**5021**	MNC496W	**5024**	GBU24V	**5028**	GBU28V	**5030**	MNC499W
5019	MNC495W	**5022**	GBU22V	**5026**	GBU26V				

5036-5110		MCW Metrobus DR102/21		MCW		H43/30F	1981		
5036	MRJ36W	**5046**	MRJ46W	**5066**	MRJ66W	**5078**	ORJ78W	**5095**	ORJ95W
5037	MRJ37W	**5047**	MRJ47W	**5067**	MRJ67W	**5079**	ORJ79W	**5098**	ORJ98W
5038	MRJ38W	**5048**	MRJ48W	**5071**	MRJ71W	**5080**	ORJ80W	**5100**	ORJ100W
5040	MRJ40W	**5049**	MRJ49W	**5072**	ORJ72W	**5081**	ORJ81W	**5106**	SND106X
5041	MRJ41W	**5051**	MRJ51W	**5073**	ORJ73W	**5091**	ORJ91W	**5107**	SND107X
5042	MRJ42W	**5052**	MRJ52W	**5074**	ORJ74W	**5092**	ORJ92W	**5108**	SND108X
5043	MRJ43W	**5053**	MRJ53W	**5075**	ORJ75W	**5093**	ORJ93W	**5109**	SND109X
5044	MRJ44W	**5054**	MRJ54W	**5076**	ORJ76W	**5094**	ORJ94W	**5110**	SND110X
5045	MRJ45W	**5055**	MRJ55W	**5077**	ORJ77W				

5111-5190		MCW Metrobus DR102/23		MCW		H43/30F	1981-83		
5111	SND111X	**5123**	SND123X	**5144**	SND144X	**5159**	ANA159Y	**5165**	ANA165Y
5116	SND116X	**5124**	SND124X	**5145**	SND145X	**5160**	ANA160Y	**5170**	ANA170Y
5117	SND117X	**5125**	SND125X	**5153**	ANA153Y	**5161**	ANA161Y	**5173**	ANA173Y
5118	SND118X	**5132**	SND132X	**5154**	ANA154Y	**5162**	ANA162Y	**5179**	ANA179Y
5119	SND119X	**5141**	SND141X	**5155**	ANA155Y	**5163**	ANA163Y	**5180**	ANA180Y
5120	SND120X	**5142**	SND142X	**5157**	ANA157Y	**5164**	ANA164Y	**5190**	ANA190Y
5121	SND121X	**5143**	SND143X	**5158**	ANA158Y				

7032	VNB132L	Leyland Atlantean AN68/1R	Park Royal	O43/32F	1972
7564	JND998N	Leyland Atlantean AN68A/1R	Northern Counties	H43/32F	1975
7687	ONF687R	Leyland Atlantean AN68A/1R	Northern Counties	H43/32F	1977
7696	ONF696R	Leyland Atlantean AN68A/1R	Northern Counties	H43/32F	1977

Recently repainted, and awaiting new vinyls, is 7729, RJA729R. It was photographed in Stockport while on lay-over. At the foot of the windscreen are the stickers used to differentiate the GMS Buses' vehicles from those of GMN. *Andrew Jarosz*

7701-7778 Leyland Atlantean AN68A/1R Northern Counties H43/32F 1977

7701	RJA701R	**7729**	RJA729R	**7759**	SRJ759R	**7766**	UNA766S	**7778**	UNA778S
7718	RJA718R	**7736**	SRJ736R	**7748**	SRJ748R				

7813-7960 Leyland Atlantean AN68A/1R Park Royal H43/32F 1977-79

7804	RJA804R	**7856**	UNA856S	**7889**	WVM889S	**7912**	ANC912T	**7944**	BNC944T
7813	RJA813R	**7860**	UNA860S	**7891**	WVM891S	**7913**	ANC913T	**7945**	BNC945T
7814	RJA814R	**7862**	UNA862S	**7892**	WVM892S	**7915**	ANC915T	**7948**	BNC948T
7815	RJA815R	**7868**	UNA868S	**7894**	WVM894S	**7919**	ANC919T	**7953**	BNC953T
7820	UNA820S	**7870**	WVM870S	**7895**	WVM895S	**7924**	ANC924T	**7954**	BNC954T
7831	UNA831S	**7872**	WVM872S	**7896**	WVM896S	**7925**	ANC925T	**7955**	BNC955T
7836	UNA836S	**7878**	WVM878S	**7899**	WVM899S	**7926**	ANC926T	**7956**	BNC956T
7838	UNA838S	**7879**	WVM879S	**7900**	WVM900S	**7929**	BNC929T	**7958**	BNC958T
7843	UNA843S	**7881**	WVM881S	**7903**	ANC903T	**7939**	BNC939T	**7959**	BNC959T
7845	UNA845S	**7885**	WVM885S	**7908**	ANC908T	**7941**	BNC941T	**7960**	BNC960T
7850	UNA850S								

Previous Registrations:

476CEL	HNE642N	OIW1317	F853JVR	OXK373	A33KBA
515VTB	From new	OIW1318	F33HGG	PXI7915	E574UHS
583TD	KGS491Y	OIW1319	F35HGG	PXI8916	E578UHS
IIL1047	C707END	OJI9451	B368VBA	SIB1361	C311ENA
L543YUS	XIA257, KSK954	OJI9452	B369VBA	SIB1832	C312ENA
NIW1673	B370VBA	OJI9455	C167ANA	SIB2014	C106AFX
NIW1676	C706END	OTK802	A32KBA	TPX884	B371VBA
NIW2399	C309ENA				

Livery: Orange, white and brown; turquoise blue (Birkenhead & District); pink, orange, red, brown and white (Express) 3255/60/8/9/72/7, 3139/98, 3213/4

Charterplan: 1-100

HAPPY AL's

T A Cullinan, 23 Lingham Lane, Moreton, Wirral, Merseyside, L46 7SA

1	HTU164N	Bristol VRT/SL2/6LX	Eastern Coach Works	H43/31F	1975	Ex Crosville, 1987
3	OPL216R	Bristol VRT/SL3/6LXB	Eastern Coach Works	DPH41/29F	1977	Ex Avon Buses, Moreton, 1992
4	MOU745R	Bristol VRT/SL3/6LXB	Eastern Coach Works	DPH39/28F	1977	Ex Badgerline, 1992
5	MOU741R	Bristol VRT/SL3/6LXB	Eastern Coach Works	DPH39/28F	1977	Ex Badgerline, 1992
6	CBV15S	Bristol VRT/SL3/501	Eastern Coach Works	H43/31F	1978	Ex Southend, 1992
7	TWS907T	Bristol VRT/SL3/6LXB	Eastern Coach Works	DPH39/28F	1979	Ex Badgerline, 1992
8	AHW204V	Bristol VRT/SL3/6LXB	Eastern Coach Works	H43/30F	1980	Ex City Line, 1993
9	AHW205V	Bristol VRT/SL3/6LXB	Eastern Coach Works	H43/30F	1980	Ex City Line, 1993
10	AHU518V	Bristol VRT/SL3/6LXB	Eastern Coach Works	H43/30F	1980	Ex City Line, 1993
11	AHU520V	Bristol VRT/SL3/6LXB	Eastern Coach Works	H43/30F	1980	Ex City Line, 1993
12	AHU521V	Bristol VRT/SL3/6LXB	Eastern Coach Works	H43/30F	1980	Ex City Line, 1993
13	AHU522V	Bristol VRT/SL3/6LXB	Eastern Coach Works	H43/30F	1980	Ex City Line, 1993
14	ALS43Y	Leyland Olympian ONTL11/2R	Eastern Coach Works	CH45/28F	1984	Ex Avon Buses, Moreton, 1992
15	A13ALS	Leyland Olympian ONTL11/2RH	Eastern Coach Works	CH45/28F	1985	Ex Thamesway, 1991
16	A14ALS	Leyland Olympian ONTL11/2RH	Eastern Coach Works	CH45/28F	1985	Ex Thamesway, 1991
17	A9ALS	Leyland Olympian ONLXB/1R	Eastern Coach Works	DPH42/29F	1986	Ex Crosville Wales, 1990
18	A16ALS	Leyland Olympian ONLXB/1R	Eastern Coach Works	DPH42/29F	1986	Ex Crosville Wales, 1990
19	A17ALS	Leyland Olympian ONLXB/1R	Eastern Coach Works	DPH42/29F	1986	Ex Crosville Wales, 1990
20	KJO503W	Bristol VRT/SL3/6LXB	Eastern Coach Works	H43/27D	1980	Ex Oxford Bus Company, 1993
21	KJO506W	Bristol VRT/SL3/6LXB	Eastern Coach Works	H43/27D	1980	Ex Oxford Bus Company, 1993
22	KJO508W	Bristol VRT/SL3/6LXB	Eastern Coach Works	H43/27D	1980	Ex Oxford Bus Company, 1993
23	KJO509W	Bristol VRT/SL3/6LXB	Eastern Coach Works	H43/27D	1980	Ex Oxford Bus Company, 1993
24	PFC511W	Bristol VRT/SL3/6LXB	Eastern Coach Works	H43/27D	1981	Ex Oxford Bus Company, 1993
25	PFC513W	Bristol VRT/SL3/6LXB	Eastern Coach Works	H43/27D	1981	Ex Oxford Bus Company, 1993

New to National Welsh as D627YCX, and arriving with Happy Al's from Tellings-Golden Miller, AKG231A, is a DAF mid-engined coach with Plaxton Paramount 3200 Mark III bodywork. Now re-registered to A19ALS, it was photographed in tree-lined Exhibition Road, South Kensington.
Keith Grimes

Happy Al's now operate six Leyland Olympians, often found on Merseytravel Bus services. Seen heading for Heswall is 17, A9ALS, one of three originating with Crosville Wales. *Simon Watts*

CRY33T	Bedford YMT	Plaxton Supreme III	C53F	1978	Ex Boyden, Castle Donington, 1987
KSD91W	Volvo-Ailsa B55-10	Alexander AV	H44/35F	1980	Ex Clydeside 2000, 1993
KSD97W	Volvo-Ailsa B55-10	Alexander AV	H44/35F	1980	Ex Clydeside 2000, 1993
A12ALS	Bova FHD12.280	Bova Futura	C45FT	1985	Ex Crosville Wales, 1990
A18ALS	Bedford YNT	Plaxton Paramount 3200	C57F	1985	Ex Ruxley, Tolworth, 1988
A19ALS	DAF MB230DKFL615	Plaxton Paramount 3200 III	C51FT	1987	Ex Tellings - Golden Miller, Cardiff, 1992
A20ALS	DAF MB230DKFL615	Plaxton Paramount 3200 III	C51FT	1987	Ex Tellings - Golden Miller, Cardiff, 1992
A10ALS	Leyland Royal Tiger RTC	Leyland Doyen	C46FT	1987	Ex West Riding, 1989
IIL4012	Leyland Royal Tiger RTC	Leyland Doyen	C46FT	1987	Ex Lancaster, 1993
PJI3749	Leyland Royal Tiger RTC	Leyland Doyen	C49FT	1987	Ex Sinclair, Greenhead, 1993
E444MMM	Van Hool T815H	Van Hool Acron	C53FT	1988	Ex Merlyns, Skewen, 1993
E91VWA	Aüwaerter Neoplan N122	Aüwaerter Skyliner	CH57/20CT	1987	Ex Express Travel, Perth, 1994
F48ALS	DAF SB3000DHS585	Van Hool Alizée	C51FT	1989	Ex Hurst & Leak, Goose Green, 1993

Previous Registrations:

A10ALS	E49TYG	A17ALS	C207GTU	ALS34Y	YPJ504Y
A12ALS	B59DMB	A18ALS	B448CMC		
A13ALS	B692BPU	A19ALS	D627YCX, AKG231A	F48ALS	F272RJX
A14ALS	B695BPU	A20ALS	D625YCX, AAX528A	IIL4012	D457EEG
A16ALS	C206GTU	A9ALS	C205GTU	PJI3749	E44MMT

Livery: White, red, orange and yellow.

HEATONS TRAVEL

H & CE Heaton, Holden Road, Leigh, Greater Manchester, WN7 1JN

UPE204M	Leyland National 1051/1R		B41F	1974	Ex Red & Green, Chislehurst, 1994
LPB194P	Leyland National 10351/1R		DP39F	1976	Ex London & Country, 1992
YFR497R	Leyland Leopard PSU3E/4R	Duple Dominant I	C47F	1977	Ex Hyndburn, 1992
KRN101T	Leyland Leopard PSU3E/4R	Duple Dominant II Express	C47F	1978	Ex Hyndburn, 1993
GSX866T	Leyland National 11351A/1R		B52F	1978	Ex Victoria Travel, Earlestown, 1994
AAK105T	Leyland National 10351B/1R		B44F	1979	Ex The Wright Company, Wrexham, 1994
AAK107T	Leyland National 10351B/1R		B44F	1979	Ex The Wright Company, Wrexham, 1994
CNH174X	Leyland Leopard PSU3G/4RT	Eastern Coach Works B51	C49F	1982	Ex United Counties, 1993
ACH941A	DAF MB200DKTL600	Caetano Alpha	C51F	1981	Ex Gray, Wigan, 1994
UXI8637	DAF MB200DKFL600	Duple Caribbean	C49FT	1984	Ex Pennine Blue, Dukinfield, 1994
B472WTC	Ford Transit 190	Robin Hood	B16F	1988	Ex Eastern National, 1994
C457BHY	Ford Transit 190	Dormobile	B16F	1989	Ex Eastern National, 1994
C551BHY	Ford Transit 190	Dormobile	B16F	1989	Ex Badgerline, 1994
C553BHY	Ford Transit 190	Dormobile	B16F	1989	Ex Badgerline, 1994
C563BHY	Ford Transit 190	Dormobile	B16F	1989	Ex Badgerline, 1994
C564BHY	Ford Transit 190	Dormobile	B16F	1989	Ex Badgerline, 1994
C567BHY	Ford Transit 190	Dormobile	B16F	1989	Ex Badgerline, 1994
D900MDB	Renault-Dodge S56	Northern Counties	B20F	1987	Ex Victoria Travel, Earlestown, 1993
E575ANE	Renault-Dodge S56	Northern Counties	B20F	1988	
F139HNC	Renault-Dodge S56	Northern Counties	B23F	1988	
F140HNC	Renault-Dodge S56	Northern Counties	B23F	1988	
K167FYG	Optare MetroRider	Optare	B31F	1992	
K168FYG	Optare MetroRider	Optare	B31F	1992	

Livery: Red and white

Previous Registrations:

ACH941A	UFP955X	UXI8637	B703MHN

The MCW MetroRider has recently been re-launched with the Cummins Euro 1 specification engine. Some 1500 MetroRiders are now in service, and Heatons Travel added two in 1992. K167FYG is seen in Wigan Bus Station.

HULME HALL

Hulme Hall Coaches Ltd, 75 Hulme Hall Road, Cheadle Hulme,
Greater Manchester, SK8 6JZ

PEC345V	Ford R1114	Plaxton Supreme IV	C53F	1980	Ex Bibby, Ingleton, 1987
RMA432V	Bristol VRT/SL3/501	Eastern Coach Works	H43/31F	1980	Ex Crosville Wales, 1991
PFC514W	Bristol VRT/SL3/6LXB	Eastern Coach Works	H43/27D	1981	Ex Oxford Bus Company, 1993
PFC515W	Bristol VRT/SL3/6LXB	Eastern Coach Works	H43/27D	1981	Ex Oxford Bus Company, 1993
WTU490W	Bristol VRT/SL3/501	Eastern Coach Works	H43/31F	1981	Ex PMT, 1990
SND288X	Leyland Leopard PSU5D/4R	Duple Dominant IV	C53F	1982	Ex North Western, 1990
RJI4082	MCW Metroliner DR130/3	MCW	CH57/25FT	1984	Ex North Devon, 1992
C360BRJ	Freight Rover Sherpa	Dixon Lomas	C16F	1985	
RJI8918	Leyland Tiger TRCTL11/3R	Duple 340	C49FT	1986	Ex ?, 1994
RJI4083	MCW Metroliner DR130/23	MCW	CH57/18FT	1986	Ex Happy Days, Woodseaves, 1991
CSU917	MCW Hiliner HR131/11	MCW	C48FT	1987	Ex Central Coachways, 1992

Previous Registrations:

CSU917	D905POF	RJI4083	C647FTT
RJI4082	A762VAF	RJI8918	C75KLG

Hulme Hall operate six Bristol VRs, four being standard bus versions. Resplendent after a repaint is WTU490W, seen entering the Middle Walk in Blackpool on the occasion of the Ribble Enthusiast's annual rally between Southport and Blackpool. *Simon Watts*

HYNDBURN

Hyndburn Transport Ltd, Ellison Street, Accrington, Lancashire, BB5 1RB

1	MIW1608	Toyota Coaster HB31R	Caetano Optimo	C19F	1987	Ex Mandale, Penrith, 1991
2	MIW9048	Bova EL26/581	Bova Europa	C47F	1982	Ex Crescent, North Walsham, 1992
3	J263KRN	Leyland Swift ST2R44C97A4	Reeve Burgess Harrier	B39F	1991	
4	J264KRN	Leyland Swift ST2R44C97A4	Reeve Burgess Harrier	B39F	1991	
5	FIL7287	Bova EL26/581	Bova Europa	C49F	1983	Ex Crescent, North Walsham, 1992
6	3402FM	Volvo B10M-61	Van Hool Alizée	C46FT	1983	Ex Fraser, Accrington, 1993
19	L448FFR	Iveco TurboDaily 59.12	Mellor	B27F	1994	
20	L447FFR	Iveco TurboDaily 59.12	Mellor	B27F	1994	
21	L445FFR	Iveco TurboDaily 59.12	Mellor	B27F	1994	
22	L446FFR	Iveco TurboDaily 59.12	Mellor	B27F	1994	
23	D860FOT	Iveco Daily 49.10	Robin Hood City Nippy	B19F	1987	Ex Ribble, 1993
24	E24GCK	MCW MetroRider MF151/6	MCW	B23F	1987	
25	E25GCK	MCW MetroRider MF151/6	MCW	B23F	1987	
26	E26GCK	MCW MetroRider MF151/6	MCW	B23F	1987	
27	E27GCK	MCW MetroRider MF151/7	MCW	DP25F	1987	
28	E28GCK	MCW MetroRider MF151/7	MCW	DP25F	1987	
29	E29GCK	MCW MetroRider MF151/7	MCW	DP25F	1987	
30	F913HTU	Iveco Daily 49-10	Robin Hood City Nippy	B25F	1989	
31	F914HTU	Iveco Daily 49-10	Robin Hood City Nippy	B25F	1989	
32	F882CJC	Iveco Daily 49-10	Robin Hood City Nippy	B25F	1989	Ex Alpine, Llandudno, 1989
33	F883CJC	Iveco Daily 49-10	Robin Hood City Nippy	B25F	1989	Ex Alpine, Llandudno, 1989
34	G41XBK	Iveco Daily 49-10	Phoenix	B25F	1990	
35	G767CDU	Leyland Swift LBM6T/2RA	Reeve Burgess Harrier	B39F	1990	Ex Volvo demonstrator, 1990
36	H36YCW	Leyland Swift ST2R44C97A4	Reeve Burgess Harrier	B39F	1990	
37	H37YCW	Leyland Swift ST2R44C97A4	Reeve Burgess Harrier	B39F	1990	
38	H38YCW	Leyland Swift ST2R44C97A4	Reeve Burgess Harrier	B39F	1990	
39	H39YCW	Leyland Swift ST2R44C97A4	Reeve Burgess Harrier	B39F	1990	
41	JFV294N	Leyland Leopard PSU3B/4R	East Lancashire(1983)	B52F	1975	Ex Halton, 1988
42	JFV295N	Leyland Leopard PSU3B/4R	East Lancashire(1983)	B52F	1975	Ex Halton, 1988

43-48 Leyland Leopard PSU4E/2R | East Lancashire | DP43F | 1977-78

43	EHG43S	**45**	EHG45S	**46**	EHG46S	**47**	NFR747T	**48**	NFR748T
44	EHG44S								

49	GCK49W	Leyland Leopard PSU4E/2R	East Lancashire	DP43F	1981	
50	A50LHG	Dennis Falcon SDA413	East Lancashire	DP42F	1984	
51	B51XFV	Dennis Falcon SDA413	East Lancashire	DP40F	1985	
52	AFM1W	Leyland National 2 NL116AL11/2R	East Lancs Greenway(1992)	B48F	1981	Ex North Western, 1992
54	KIB7257	Leyland Leopard PSU3E/4R	Duple Dominant II	C49F	1978	Ex PG Travel, Middlewich, 1978
55	VNH157W	Leyland Leopard PSU3F/4R	Duple Dominant IV Express	C49F	1981	Ex Milton Keynes, 1987
56	YKA8W	Leyland Leopard PSU3F/4R	Duple Dominant I	C49F	1981	Ex Merseybus, 1988
57	WWM576W	Leyland Leopard PSU3F/4R	Duple Dominant I	C49F	1980	Ex Merseybus, 1988
62	YFR492R	Leyland Leopard PSU3E/4R	Duple Dominant I	C47F	1977	Ex Ribble, 1988
63	YFR493R	Leyland Leopard PSU3E/4R	Duple Dominant I	C47F	1977	Ex Ribble, 1988
66	YFR496R	Leyland Leopard PSU3E/4R	Duple Dominant I	C47F	1977	Ex Ribble, 1988
68	BKJ153T	Leyland Leopard PSU3E/4R	Duple Dominant I	C49F	1979	Ex Maidstone & District, 1990
70	BKJ150T	Leyland Leopard PSU3E/4R	Duple Dominant I	C49F	1979	Ex Maidstone & District, 1990

75-79 Leyland Leopard PSU5C/4R | Duple Dominant II | C57F | 1980 | Ex Maidstone & District, 1990

75	MNK421V	**76**	MNK422V	**77**	MNK423V	**78**	MNK424V	**79**	MNK425V

134	JKW286W	Leyland Atlantean AN68B/1R	Alexander AL	H45/32F	1981	Ex Lodge, High Easter, 1992

135-139 Leyland Atlantean AN68C/1R | East Lancashire | H46/30F | 1980-81 | Ex Ipswich, 1991

135	RGV40W	**136**	RBJ36W	**137**	RGV37W	**138**	RGV38W	**139**	RGV39W

140-144 Leyland Atlantean AN68B/1R | Willowbrook | H45/33F | 1980 | Ex Merseybus, 1989

140	WWM920W	**141**	AFY191X	**142**	WWM922W	**143**	WWM933W	**144**	AFY181X

Above: **Originally Crosville SNL1, AFM1W was later donated to the Greenway project from North Western. After conversion it was used by East Lancashire as a demonstrator before becoming Hyndburn 52 in 1992. It is seen in Bacup, a town near the Yorkshire boundry and now part of Rossendale district.** *Ken Grime*

Opposite, top: **Seven of the short Leyland Leopard, the PSU4 variant were supplied to Hyndburn between 1977 and 1981, all fitted with high-back seating for use on the longer rural services radiating from Accrington. Seen recently in Accrington bus station is 43, EHG43S.** *Phillip Stephenson*

Opposite, bottom: **Newly arrived with Hyndburn is a quartet of Iveco TurboDaily 59.12s bodied by Mellor Coachcraft, some ten miles away in Rochdale. Seen heading for Baxendale on service 420 is 21, L445FFR.** *Andrew Jarosz*

194	KHG194T	Leyland Atlantean AN68A/1R	East Lancashire	H45/33F	1978	
196	VCW196V	Leyland Atlantean AN68A/1R	East Lancashire	H45/33F	1979	
197	VCW197V	Leyland Atlantean AN68A/1R	East Lancashire	H45/33F	1979	
199	DBV199W	Leyland Atlantean AN68B/1R	East Lancashire	H45/33F	1980	
208	NRN383P	Leyland Atlantean AN68/1R	Park Royal	H43/30F	1976	Ex Ribble, 1990
209	GBV109N	Leyland Atlantean AN68/2R	Alexander AL	H49/35F	1975	Ex Preston, 1990
210	GBV110N	Leyland Atlantean AN68/2R	Alexander AL	H49/35F	1975	Ex Preston, 1990
211	GBV101N	Leyland Atlantean AN68/2R	Alexander AL	H49/35F	1975	Ex Preston, 1990
212	GBV108N	Leyland Atlantean AN68/2R	Alexander AL	H49/35F	1975	Ex Preston, 1990
214	NRN390P	Leyland Atlantean AN68/1R	Park Royal	H43/30F	1976	Ex Ribble, 1990
215	NRN402P	Leyland Atlantean AN68/1R	Park Royal	H43/30F	1976	Ex Ribble, 1990
216	NRN404P	Leyland Atlantean AN68/1R	Park Royal	H43/30F	1976	Ex Ribble, 1990
301	D42DNH	Iveco Daily 49.10	Robin Hood City Nippy	B21F	1987	Ex Rennie, Dunfermline, 1994
302	D557HNW	Iveco Daily 49.10	Robin Hood City Nippy	DP21F	1986	Ex , Sunderland, 1994
303	C210MCW	Ford Transit 190D	Carlyle	B20F	1900	
381	SVM171W	Volvo B10M-61	Jonckheere Bermuda	C49FT	1981	Ex Rigby, Patricroft, 1993
384	BGG164S	Volvo B58-61	Plaxton Supreme III	C57F	1978	Ex Rigby, Patricroft, 1993
385	HFB845X	Volvo B58-61	Duple Dominant	C53F	1981	Ex Rigby, Patricroft, 1993
386	MRJ276W	Leyland Leopard PSU5C/4R	Plaxton Supreme IV	C50F	1981	Ex Ribble, 1990
387	SWH127T	Volvo B58-61	Plaxton Supreme IV	C49FT	1979	Ex Rigby, Patricroft, 1993
389	SND289X	Leyland Leopard PSU5D/4R	Duple Dominant IV	C53F	1981	Ex Ribble, 1990
390	D755LAX	Mercedes-Benz L608D	Reeve Burgess	C19F	1986	Ex Rigby, Patricroft, 1993
395	EYN165	Leyland Tiger TRCTL11/3R	Duple Laser	C50F	1983	Ex Turner, Bury, 1993

Previous Registrations:

3402FM	PGC522Y	MIW1608	E259GKS
EYN165	BPR105Y	MIW9048	YUR828X, 3253VU, XBC140X
FIL7287	CDC879Y	SVM171W	WNV822W, 1656RU
KIB7257	LWL5S	SWH127T	CJF7T, 9874ND

Livery: Blue, red and grey; green and white (Rigbys) - 301-3, 381-395

7
FERN GORE
VIA ACCRINGTON
HYNDBURN
HYNDBURN
EHG 43S

420
BAXENDEN
NO SMOKING BUS
IVECO
L445 FFR

ISLE OF MAN

Isle of Man Department of Tourism and Transport, Terminus Buildings, Strathallan Crescent, Douglas, Isle of Man.

1	BMT401T	Leyland Lynx LX112L10ZR1R		Leyland Lynx		B51F	1989		
2	BMT402T	Leyland Lynx LX112L10ZR1R		Leyland Lynx		B51F	1989		
3-8		Dennis Dart 9SDL3016		Plaxton Pointer		B35F	1992	*9-12 are 9SDL3011	
3	CMN103L	**5**	CMN105L	**6**	CMN106L	**7**	CMN107L	**8**	CMN108L
4	CMN104L								
9-13		Dennis Dart 9SDL3011		Reeve Burgess Pointer		B35F	1992		
9	CMN9H	**10**	CMN10H	**11**	CMN11H	**12**	CMN12H	**13**	CMN13H
18	MAN18D	Leyland National 11351/1R				B52F	1975		
20	MAN20D	Leyland National 11351/1R				B52F	1975		
21-27		Leyland National 11351/1R				B52F	1976		
21	MAN21H	**23**	MAN23H	**25**	MAN25H	**26**	MAN26H	**27**	MAN27H
22	MAN22H	**24**	MAN24H						
28-33		Leyland National 11351A/1R				B52F	1977		
28	MAN28N	**30**	MAN30N	**31**	MAN31N	**32**	MAN32N	**33**	MAN33N
29	MAN29N								
34-44		Leyland Atlantean AN68A/1R		Alexander AL		H45/30F*	1978	Ex Portsmouth Transit, 1991-92 34/5/7/41/2 are H45/28D	
34	CMN34C	**37**	CMN37C	**39**	CMN39C	**41**	CMN41C	**43**	CMN43C
35	CMN35C	**38**	CMN38C	**40**	CMN40C	**42**	CMN42C	**44**	CMN44C
36	CMN36C								
46	CMN46C	Leyland Atlantean AN68A/1R		Alexander AL		H45/30F	1978	Ex Portsmouth Transit, 1992	
47	CMN47C	Leyland Atlantean AN68A/1R		Alexander AL		H45/30F	1978	Ex Portsmouth Transit, 1992	
48	CMN48C	Leyland Atlantean AN68A/1R		Alexander AL		H45/30F	1978	Ex Portsmouth Transit, 1992	
49-53		Dennis Dominator DDA1006		East Lancashire		H45/30F	1986	Ex Kingston-upon-Hull, 1993	
49	CMN49T	**50**	CMN50T	**51**	CMN51T	**52**	CMN52T	**53**	CMN52T

The Isle of Man has recently added further Dennis Darts to its fleet, displacing older Leyland Nationals in the process. One of the 1992 delivery, and fitted with Plaxton Pointer bodywork, is 5, CMN105L. It is seen in Douglas.
Geoff Mills

Port Erin lies in the south west of the Isle of Man, and 37, CMN37C, waits time before returning to Port St Mary. New to Portsmouth, it is one example of the batch to have retained its dual-door configuration.
Geoff Mills

With the Irish Sea as a background, Isle of Man 49, CMN49T, is seen while working service 1. One of five double decks to arrive on the Island recently, it is a Dennis Dominator with East Lancashire bodywork that was new to Hull.
Geoff Mills

Ninteen Leyland Olympians operate on the Isle of Man, including 65, BMN65P, an example with Northern Counties bodywork. It was photographed while heading back to Douglas.
Geoff Mills

58-64		Leyland Olympian ON2R50C16Z4		Northern Counties		H47/30F	1990		
58	BMN58V	**60**	BMN60V	**62**	BMN62V	**63**	BMN63V	**64**	BMN64V
59	BMN59V	**61**	BMN61V						

65-70		Leyland Olympian ONCL10/1RZ		Northern Counties		H47/30F	1989		
65	BMN65P	**67**	BMN67M	**68**	BMN68M	**69**	BMN69M	**70**	BMN70M
66	BMN66P								

71	CMN71P	Dennis Dart 9SDL3011	Plaxton Pointer	B35F	1993

72-79		Dennis Dart 9.8SDL3035		Plaxton Pointer		B40F	1994		
72	CMN72X	**74**	CMN74X	**76**	CMN76X	**78**	CMN78X	**79**	CMN79X
73	CMN73X	**75**	CMN75X	**77**	CMN77X				

83-87		Leyland Olympian ONCL10/1RZ		Leyland		H47/31F	1988		
83	BMN83G	**84**	BMN84G	**85**	BMN85G	**86**	BMN86G	**87**	BMN87G

88	BMN88G	Leyland Olympian ONCL10/1RZ	Leyland	DPH43/29F	1988

Previous Registrations:

2219MAN	NFN67M	CMN42C	UOR328T	CMN52T	C127CAT
CMN34C	UOR320T	CMN43C	UOR329T	CMN53T	C130CAT
CMN35C	UOR321T	CMN44C	UOR330T	MAN14A	MAN10M, MAN10A, MN9514
CMN36C	UOR322T	CMN46C	UOR332T	MAN19D	NFN62M
CMN37C	UOR323T	CMN47C	UOR333T	MAN2692	XKC858K
CMN38C	UOR324T	CMN48C	UOR334T	MAN5832	XKC835K
CMN39C	UOR325T	CMN49T	C120CAT	V648MAN	XKC819K
CMN40C	UOR326T	CMN50T	C121CAT	MAN1378	XKC789J
CMN41C	UOR327T	CMN51T	C126CAT		

Livery: Red and cream

Only two of the 1975 delivery of Leyland Nationals remain in service on the Island, most of the others having returned to the mainland for conversion to Greenways. Douglas is the setting for this picture of 18, MAN18D, as it apparently runs along the tram lines. *Geoff Mills*

KIRKBY LONSDALE MINI BUSES

S J Sutton, Twenty Acres, Moor End, Hutton Roof, Kirkby Lonsdale, Carnforth, Lancashire, LA6 2PF

DUI4760	Leyland Leopard PSU5/4R	Plaxton Supreme III	C57F	1978	Ex M&S Coaches, Forrest Gate, 1993
B609BGA	Mercedes-Benz L608D		C21F	1984	Ex ?, Carlisle, 1993
D140NUS	Mercedes-Benz L608D	Alexander AM	B21F	1986	Ex Kelvin Scottish, 1992
E460YLG	Mercedes-Benz 609D		B26F	19	Ex ?, Hereford, 19.94

Previous Registrations:
DUI4760 BGJ315S

Livery: White, maroon and blue

LITTLE WHITE BUSES

J M James, 48 Southport Road, Ormskirk, Lancashire, L39 1QR

D888MJA	Renault-Dodge S46	Northern Counties	B24F	1987	
D64MTG	Renault-Dodge S56	East Lancashire	B24F	1987	Ex Newport, 1990
D867MDB	Renault-Dodge S56	Northern Counties	B22F	1987	Ex G M Buses, 1992
D375JUM	Volkswagen LT55	Optare City Pacer	B25F	1987	Ex London Buses, 1994
E209KCK	Renault-Dodge S46	Northern Counties	B24F	1988	
E224BVR	Renault-Dodge S46	Northern Counties	B24F	1988	
E225BVR	Renault-Dodge S46	Northern Counties	B24F	1988	
E386CNE	Renault-Dodge S56	Northern Counties	B23F	1988	Ex Northern Counties demonstrator, 1989
J644LKD	Renault-Dodge S56	Northern Counties	B22F	1991	Ex Northern Counties demonstrator, 1991
K787VNR	Mercedes-Benz 811D	Wright	B33F	1993	
K457EVC	Mercedes-Benz 811D	Dormobile Routemaker	B31F	1993	
L529XNR	Mercedes-Benz 811D	Dormobile Routemaker	B31F	1993	
L193DBC	Mercedes-Benz 811D	Marshall C16	B31F	1994	
M689FJF	Mercedes-Benz 811D	Marshall C16	B31F	1994	
M	Mercedes-Benz 811D	Marshall C16	B31F	1994	On order

Livery: White and blue

On order for early 1995: 2 Mercedes-Benz 709D/Marshall C19 B27F and 1 Mercedes-Benz 811D/Marshall C16 B31F

Little White Buses operate around the Lancashire new town of Skelmersdale which has a multiplicity of roundabouts. J644LKD, a former Northern Counties demonstator for their body based upon the Renault-Dodge S56, is seen in the bus station while working service 10.
Mark Bailey

LIVERBUS

Gemsam Ltd, New Garage, Wilson Road, Huyton, Merseyside, L36 7AH
London Suburban District Bus Co Ltd, Gemsam Buildings, Wilson Road, Huyton, Merseyside, L36 6JG

Depots: Wilson Road, Huyton; Hastingwood Trading Estate, Edmonton, London.

10	A105HNC	Leyland Tiger TRCTL11/3R	Plaxton Paramount 3200	C46FT	1984	Ex Crosville Wales, 1990
14	J531JNH	Volvo B10M-60	Jonckheere Deauville P599	C51FT	1992	
18	B693BPU	Leyland Olympian ONTL11/2RH	Eastern Coach Works	CH45/28F	1985	Ex Thamesway, 1991
19	B694BPU	Leyland Olympian ONTL11/2RH	Eastern Coach Works	CH45/28F	1985	Ex Thamesway, 1991

20-38 Leyland Atlantean AN68A/1R Northern Counties H43/32F 1976-77 Ex GM Buses, 1990

20	ONF661R	**24**	ONF674R	**28**	ONF694R	**32**	SRJ732R	**36**	SRJ753R
21	ONF662R	**25**	ONF675R	**29**	RJA704R	**33**	SRJ739R	**37**	SRJ754R
22w	ONF663R	**26**	ONF678R	**30**	RJA713R	**34**	SRJ749R	**38**	SRJ758R
23	ONF668R	**27**	ONF690R	**31**	SRJ731R	**35**	SRJ752R		

39-50 Leyland Atlantean AN68A/1R Northern Counties H43/32F 1976-77 Ex GM Buses, 1990

39	ONF658R	**42**	ONF689R	**44**	RJA723R	**47**	UNA770S	**49**	SRJ742R
40	ONF670R	**43**	ONF693R	**46**	UNA763S	**48**	UNA776S	**50**	RJA727R
41	ONF671R								

Liverbus took delivery of the first Northern Counties-bodied Volvo B10Bs in 1993, the bodywork being to the Dutch-inspired Paladin design. Seen heading for Liverpool's Pier Head is 103, K103OHF.
Roy Marshall

52	UNA851S	Leyland Atlantean AN68A/1R	Park Royal	H43/32F	1977	Ex GM Buses, 1990
53	UNA788S	Leyland Atlantean AN68A/1R	Northern Counties	H43/32F	1977	Ex GM Buses, 1990
55	BNC935T	Leyland Atlantean AN68A/1R	Park Royal	H43/32F	1977	Ex GM Buses, 1990
60	C375CAS	Leyland Olympian ONLXB/1R	Alexander RL	DPH45/27F	1984	Ex Highland Scottish, 1992
61	C376CAS	Leyland Olympian ONLXB/1R	Alexander RL	DPH45/27F	1984	Ex Highland Scottish, 1992
62	C377CAS	Leyland Olympian ONLXB/1R	Alexander RL	H45/27F	1984	Ex Highland Scottish, 1992
80	XAK459T	Leyland National 11351A/1R		B52F	1978	Ex Rye Coaches, 1994
81	MNH568V	Leyland National 11351A/1R		B49F	1979	Ex Brylaine, Boston, 1994
82	PJI5912	Leyland National 11351A/1R		B49F	1978	Ex Beeston, Hadleigh, 1994
83	PJI5913	Leyland National 11351A/1R		B49F	1978	Ex Beeston, Hadleigh, 1994

101-108 Volvo B10B-58 — Northern Counties Paladin — B51F — 1993

101	K101OHF	**103**	K103OHF	**105**	K105OHF	**107**	K107OHF	**108**	K108OHF
102	K102OHF	**104**	K104OHF	**106**	K106OHF				

201-210 Volvo Olympian YN2RV18Z4 — Northern Counties Palatine II — H47/25D — 1993

201	L201SKD	**203**	L203SKD	**205**	L205SKD	**207**	L207SKD	**209**	L209SKD
202	L202SKD	**204**	L204SKD	**206**	L206SKD	**208**	L208SKD	**210**	L210SKD

211 L211SBG — Volvo Olympian YN2RV18Z4 — Northern Counties Palatine II — H47/29F — 1993

212-217 Volvo Olympian YN2RV18Z4 — Northern Counties Palatine II — H47/25D — 1994

212	L212TWM	**214**	L214TWM	**215**	L215TWM	**216**	L216TWM	**217**	L217TWM
213	L213TWM								

1501-1520 Leyland Titan TNLXB2RRSp — Park Royal — H44/25D — 1979-82 — Ex London Buses, 1993

1501	WYV61T	**1505**	CUL182V	**1509**	CUL161V	**1513**	KYN290X	**1517**	KYV538X
1502	CUL82V	**1506**	CUL77V	**1510**	CUL207V	**1514**	CUL210V	**1518**	CUL205V
1503	CUL141V	**1507**	CUL102V	**1511**	CUL178V	**1515**	CUL129V	**1519**	KYV436X
1504	CUL151V	**1508**	CUL111V	**1512**	CUL97V	**1516**	CUL170V	**1520**	KYV393X

Previous Registrations:

PJI5912 RKA872T   PJI5913 RKA874T

Livery: Brown, white and red.

Operations: London Suburban Bus - 201-10/2-7, 1501-20; Liverline remainder

Liverbus operate their former-GM Buses Leyland Atlanteans on Merseyside, with few making appearances in London. Working service 212 at Huyton is 35, SRJ752R.
Mark Bailey

Many of the Leyand Titans operated by Gemsam work on the London Suburban Bus services in London. However, vehicles do interchange, and Titan 1506, CUL77V, is seen arriving at Pier Head on service 212, its destination already turned to show the return journey. *Ken Grime*

LIVERPOOL CITY COACHES

J C Bleasdale, 99-103, Stanhope Street, Liverpool, Merseyside, L8 5RE

GNJ576N	Bristol VRT/SL2/6LX	Eastern Coach Works	H43/31F	1974	Ex Brighton & Hove, 1990
JJG908P	Leyland National 10351/1R		B41F	1976	Ex East Kent, 1989
EMB370S	Leyland National 11351A/1R		B49F	1978	Ex Crosville Wales, 1991
CWX657T	Leyland National 11351A/1R		B49F	1979	Ex West Riding, 1991
EON830V	Leyland National 2 NL116L11/1R		B47FL	1980	Ex Stevensons, 1992
LUA325V	Leyland National 2 NL106L11/1R		B41F	1980	Ex Stevensons, 1993
LCW731W	Leyland Leopard PSU3E/4R	Plaxton Supreme IV	C50F	1981	Ex Burnley & Pendle, 1994
917MMB	Leyland Tiger TRCTL11/3R	Plaxton Supreme V	C53F	1982	Ex Robinsons, Gt Harwood, 1990
D32MWN	Leyland Lynx LX2R11C15Z4S	Leyland Lynx	B49F	1990	Ex Whitelaw, Stonehouse, 1993
D390SGS	Freight Rover Sherpa	Dormobile	B16F	1987	Ex Eagles & Crawford, Mold, 1992
K782VNF	Ford Transit VE6	Deansgate	M14	1992	

Previous Registrations:
917MMB LEC198X LCW731W LHE253W, XSU910

Livery: Blue and white.

MANCHESTER INTERNATIONAL AIRPORT

Manchester Airport plc, Manchester International Airport, Ringway, Manchester, M22 5PA

N1-N8 Leyland National 2 NL116AL11/3R B30T 1981

N1	NHH358W	**N3**	Q580GRJ	**N5**	unreg	**N7**	unreg	**N8**	unreg
N2	NHH359W	**N4**	unreg	**N6**	unreg				

N9	YCD78T	Leyland National 11351A/2R		B44D	1979	Ex Southdown, 1992
N10	YCD88T	Leyland National 11351A/2R		B44D	1979	Ex Southdown, 1992
N11	WYJ164S	Leyland National 11351A/2R		B44D	1978	Ex Southdown, 1992
N12	UFG62S	Leyland National 11351A/2R		B44D	1978	Ex Southdown, 1992
N13	RUF44R	Leyland National 11351A/2R		B44D	1977	Ex Southdown, 1992
N14	XNK206X	Ford R1014	Plaxton Bustler	B29D	1982	Ex Miller, Foxton, 1992
N15	XNK209X	Ford R1014	Plaxton Bustler	B29D	1982	Ex Miller, Foxton, 1992
N16	XNK212X	Ford R1014	Plaxton Bustler	B29D	1982	Ex Miller, Foxton, 1992
N17	XNK215X	Ford R1014	Plaxton Bustler	B29D	1982	Ex Miller, Foxton, 1992
	C824CBU	Renault-Dodge S56	Northern Counties	B18F	1987	Ex GM Buses, 1990
	D50RJA	Iveco Daily 49.10	Robin Hood City Nippy	B18FL	1987	
	D51RJA	Iveco Daily 49.10	Robin Hood City Nippy	B18FL	1987	
	D52RJA	Iveco Daily 49.10	Robin Hood City Nippy	B18F	1987	
	D53RJA	Iveco Daily 49.10	Robin Hood City Nippy	B18F	1987	
	D54RJA	Iveco Daily 49.10	Robin Hood City Nippy	B18F	1987	
	E823YBA	Iveco Daily 49.10	Robin Hood City Nippy	B18F	1988	
	E824YBA	Iveco Daily 49.10	Robin Hood City Nippy	B18F	1988	
	H199ENC	Mercedes-Benz 709D	Reeve Burgess Beaver	B20F	1990	
	H201ENC	Mercedes-Benz 709D	Reeve Burgess Beaver	B20F	1990	
	H726LOL	Mercedes-Benz 811D	Carlyle	B27F	1990	
	K125TCP	DAF SB220LC550	Ikarus CitiBus	B26DL	1993	
	K126TCP	DAF SB220LC550	Ikarus CitiBus	B26DL	1993	
	K127TCP	DAF SB220LC550	Ikarus CitiBus	B26DL	1993	
	K976JWW	Optare MetroRider MR39	Optare	B23F	1993	
	K977JWW	Optare MetroRider MR39	Optare	B23F	1993	
	K978JWW	Optare MetroRider MR39	Optare	B23F	1993	

Livery: Grey, yellow and white

Manchester International Airport operate their own services within the airport, both land-side and air-side. The Leyland Nationals are confined to carrying passengers between the terminal and aircraft, while the Ikarus CitiBus and Optares are used on services between the two terminals and the new rail station. Seen outside this new facility is K127TCP.
Simon Watts

MANCHESTER METROLINK

Greater Manchester Metro Ltd, Metrolink House, Queens Road,
Manchester, M8 7RY

1001-1026		GEC Alsthom		Firema		86-seat	1991-92
1001	**1005**	**1009**	**1012**	**1015**	**1018**	**1021**	**1024**
1002	**1006**	**1010**	**1013**	**1016**	**1019**	**1022**	**1025**
1003	**1007**	**1011**	**1014**	**1017**	**1020**	**1023**	**1026**
1004	**1008**						

Livery: Ivory and turquoise

Named trams:
1002 *Manchester Arndale Voyager;* 1003 *The Robert Owen;* 1005 *Greater Altrincham Enterprise;* 1007 *The Guiness Record Breaker;* 1009 *Co-operative Insurance;* 1010 *Manchester Champion;* 1014 *Manchester 2000;* 1015 *Sparky;* 1016 *Signal Express;* 1017 *Rosie;* 1019 *The Eric Black;* 1024 *The John Greenwood;* 1025 *Christie Metro Challenger.*

The Metrolink continues to attract attention, and more passengers, as its considers its expansion plans, with extensions to Trafford Park, Oldham and Glossop on the horizon. Photographed while heading south near the G-Mex complex is 1009. At peak times twin cars may now be seen in the city.
Roy Marshall

MAYNE

A Mayne and Son Ltd, Ashton New Road, Clayton, Manchester, M11 4PD

2	THX515S	Leyland Fleetline FE30ALR Sp	Park Royal	H44/31F	1978	Ex London Buses, 1991
3	OJD163R	Leyland Fleetline FE30AGR	Park Royal	H44/29F	1976	Ex Stevensons, 1985
5	OUC35R	Leyland Fleetline FE30AGR	MCW	H45/32F	1976	Ex Stevensons, 1985
6	ULS666T	Leyland Fleetline FE30AGR	Eastern Coach Works	H43/32F	1979	Ex Clydeside 2000, 1992
7	THX322S	Leyland Fleetline FE30ALR Sp	MCW	H44/31F	1978	Ex London Buses, 1991
9	THX579S	Leyland Fleetline FE30ALR Sp	Park Royal	H44/27F	1978	Ex London Buses, 1991
10	M210NDB	Scania N113DRB	East Lancashire	H47/31F	1994	
11	M211NDB	Scania N113DRB	East Lancashire	H47/31F	1994	
12	F112HNC	Scania N113DRB	Northern Counties	H47/32F	1989	
13	F113HNC	Scania N113DRB	Northern Counties	DPH47/30F	1989	
14	L114DNA	Scania N113DRB	East Lancashire	H47/31F	1993	
15	G115SBA	Scania N113DRB	Northern Counties	H47/32F	1989	
16	G116SBA	Scania N113DRB	Northern Counties	H47/32F	1989	
17	G117SBA	Scania N113DRB	Northern Counties	H47/32F	1989	
19	THX619S	Leyland Fleetline FE30ALR Sp	MCW	H44/31F	1978	Ex London Buses, 1991
21	A101DPB	Dennis Falcon HC SD407	Wadham Stringer(1987)	DP49F	1983	Ex Wycombe Bus, 1991
22	LIW1322	Leyland Leopard PSU3B/4R	Willowbrook Warrior(1990)	B48F	1972	Ex Jowitt, Tankersley, 1990
23	LIW1323	Leyland Leopard PSU3B/4R	Willowbrook Warrior(1990)	B48F	1973	Ex Jowitt, Tankersley, 1990
25	NIB7625	Leyland Leopard PSU3C/2R	Willowbrook Warrior(1990)	B48F	1977	Ex Merthyr Tydfil, 1988
26	L26FNE	Dennis Dart 9.8SDL3035	Marshall C37	B40F	1994	
27	L27FNE	Dennis Dart 9.8SDL3035	Marshall C37	B40F	1994	
28	K28XBA	Dennis Dart 9.8SDL3012	Marshall Dartline	B40F	1992	
29	K29XBA	Dennis Dart 9.8SDL3012	Marshall Dartline	B40F	1993	
30	ULS663T	Leyland Fleetline FE30AGR	Eastern Coach Works	H43/32F	1979	Ex Clydeside 2000, 1992
31	THX601S	Leyland Fleetline FE30ALR Sp	Park Royal	H44/31F	1978	Ex London Buses, 1991
32	GSC857T	Leyland Fleetline FE30AGR	Eastern Coach Works	H43/32F	1978	Ex Clydeside 2000, 1992
33	THX303S	Leyland Fleetline FE30ALR Sp	MCW	H44/31F	1978	Ex London Buses, 1991

The former London DMS still has a role in the Mayne fleet, though investment in new buses has seen some of the older models withdrawn. Passing through Piccadilly, Manchester, is 2, THX515S.
Phillip Stephenson

Leaving Ashton-under-Lyne bus station Mayne 36, K36XNE, is a Dennis Dominator with East Lancashire bodywork. This is one of three supplied in 1993 which include DPTAC specification features for tendered bus work. Visible in this picture is the split-level entrance. *Roy Marshall*

Five Dennis Falcons with Marshall Camair bodywork were purchased from Chesterfield in 1991, and joined a Wadham Stringer-bodied example from Wycombe Bus. One of the former, 47, A47YWJ, is seen heading for Manchester. *Simon Watts*

34	THX594S	Leyland Fleetline FE30ALR Sp	Park Royal	H44/31F	1978	Ex London Buses, 1991
35	THX555S	Leyland Fleetline FE30ALR Sp	Park Royal	H44/31F	1978	Ex London Buses, 1991
36	K36XNE	Dennis Dominator DDA2005	East Lancashire	H45/31F	1993	
37	K37XNE	Dennis Dominator DDA2005	East Lancashire	H45/31F	1993	
38	K38YVM	Dennis Dominator DDA2005	East Lancashire	H45/31F	1993	
41	GDZ3841	Leyland Leopard PSU3B/4R	Willowbrook Warrior(1990)	B51F	1974	Ex Perry, Bromyard, 1990
44	A44YWJ	Dennis Falcon HC SD414	Marshall Camair	DP53F	1984	Ex Chesterfield, 1991
45	A45YWJ	Dennis Falcon HC SD414	Marshall Camair	B53F	1984	Ex Chesterfield, 1991
46	A46YWJ	Dennis Falcon HC SD414	Marshall Camair	B53F	1984	Ex Chesterfield, 1991
47	A47YWJ	Dennis Falcon HC SD414	Marshall Camair	B53F	1984	Ex Chesterfield, 1991
48	A48YWJ	Dennis Falcon HC SD414	Marshall Camair	B53F	1984	Ex Chesterfield, 1991
50	403BGO	Leyland Tiger TRCTLXCT/3RZ	Plaxton Paramount 3200	C57F	1985	
51	H51DVR	Dennis Javelin 8.5SDA1903	Duple 320	C37F	1991	
52	H52FDB	Dennis Javelin 11SDL1907	Duple 320	C55F	1991	
55	IIL1355	Leyland Tiger TRCL10/3ARZM	Duple 340	C53FT	1989	
56	IIL1356	Leyland Tiger TRCTL11/3RZA	Duple 320	C57F	1989	
57	IIL2257	Bova FHD12/290	Bova Futura	C50FT	1990	
58	IIL2258	Bova FHD12/290	Bova Futura	C50FT	1990	
59	GIL3259	Leyland Leopard PSU5C/4R	Plaxton Supreme IV	C57F	1981	
60	GIL2160	Leyland Leopard PSU5C/4R	Plaxton Supreme IV	C57F	1981	
62	NIB4162	Leyland Leopard PSU5C/4R	Plaxton Supreme IV	C57F	1982	
63	MJI5763	Leyland Leopard PSU5C/4R	Plaxton Supreme IV	C57F	1982	
64	MJI5764	Leyland Leopard PSU5C/4R	Plaxton Supreme IV	C57F	1982	
65	MJI5765	Leyland Leopard PSU5C/4R	Plaxton Supreme IV	C57F	1982	
66	MJI5766	Leyland Leopard PSU3E/4R	Plaxton Supreme IV Express	C53F	1980	
69	SSV269	Leyland Tiger TRCTL11/3R	Plaxton Paramount 3500 E	C55F	1983	
70	TKU540	Leyland Tiger TRCTL11/3R	Plaxton Paramount 3500	C49FT	1984	
77	LIB6437	Leyland Tiger TRCTLXCT/3RZ	Plaxton Paramount 3500 II	C51FT	1986	
78	LIB6438	Leyland Tiger TRCTLXCT/3RZ	Plaxton Paramount 3500 II	C51FT	1986	
79	LIB6439	Leyland Tiger TRCTLXCT/3RZ	Plaxton Paramount 3200 II	C57F	1985	

Barry Coopers Coaches

1	OJD131R	Leyland Fleetline FE30AGR	Park Royal	H44/29F	1976	Ex Stevensons, 1984
4	GND505N	Daimler Fleetline CRG6LXB	Northern Counties	H43/32F	1974	Ex Greater Manchester, 1988
8	YNA328M	Daimler Fleetline CRG6LXB	Northern Counties	H43/32F	1973	Ex Greater Manchester, 1987
10	KUC969P	Leyland Fleetline FE30ALR(6LXB)	MCW	H45/32F	1976	Ex Eilts & Dorset, 1990
17	VGU443	Leyland Leopard PSU5D/4R	Plaxton Supreme IV	C57F	1981	
18	YPL764	Leyland Tiger TRCTL11/3R	Plaxton Paramount 3500	C57F	1983	
19	YUC765	Leyland Tiger TRCTL11/3R	Duple Laser	C57F	1983	
21	EUK976	Leyland Tiger TRCTL11/3R	Plaxton Paramount 3500	C53FT	1984	
24	OED201	Leyland Tiger TRCTL11/3R	Plaxton Paramount 3200 II	C53F	1985	
25	UOL337	Leyland Tiger TRCTL11/3R	Plaxton Paramount 3200 II	C53F	1985	
26	LIB5440	Leyland Tiger TRCTLXCT/3RZ	Plaxton Paramount 3500 II	C51F	1986	
27	614BWU	DAF SB2305DHTD585	Duple 320	C57F	1989	
28	H28DVM	Dennis Javelin 11SDL1921	Duple 320	C55F	1991	
29	J29LBA	Dennis Javelin 11SDL1921	Duple 320	C55F	1991	
66	UCE665	Leyland Leopard PSU4A/4R	Plaxton Supreme IV (1982)	C53F	1968	Ex Midland Red, 1982
69	906GAU	Leyland Leopard PSU5C/4R	Plaxton Supreme IV	C57F	1980	
	DIL6907	Leyland Leopard PSU5A/4R	Duple Dominant	C57F	1976	Ex Progress, Denton, 1991
	289BUA	Leyland Leopard PSU5C/4R	Plaxton Supreme IV	C57F	1980	
	NIB3261	Leyland Leopard PSU5C/4R	Plaxton Supreme IV	C57F	1981	
	NMX643	Leyland Leopard PSU3F/5R	Plaxton Supreme IV	C53F	1980	
	WPD28Y	Leyland Leopard PSU3G/5R	Eastern Coach Works B51	C53F	1982	Ex Stagecoach South, 1994
	EHB259G	Leyland Leopard PSU4A/2R	East Lancashire	B43F	1969	Ex Stonier, Goldenhill, 1988

Previous Registrations:

289BUA	HDB356V	LIB5440	C426YBA	NIB4162	SNC362X
403BGO	B350RNA	LIB6437	C347YBA	NIB7625	NTX363R
614BWU	F27HNC	LIB6438	C348YBA	NMX643	HDB355V
906GAU	HDB357V	LIB6439	B349RNA	OED201	B424RNA
EUK978	A421KBA	LIW1322	NRE562L	SSV269	ANA368Y
GDZ3841	JVS928N	LIW1323	RBF987M	TKU540	A370HNC
GIL2160	MRJ360W	MJI5763	SNC363X	UCE665	WHA236H, SNC355X
GIL3259	MRJ359W	MJI5764	SNC364X	UOL337	B425RNA
IIL1355	F55HNC	MJI5765	SNC365X	VGU443	ODJ417W
IIL1356	F56HNC	MJI5766	MRJ358W	YPL764	A418HND
IIL2257	G57SBA	NIB3261	SNC361X	YUC785	A419HND
IIL2258	G58SBA				

MTL

MTL, Edge Lane, Liverpool, Merseyside, L7 9LL.
Fareway Passenger Services Ltd, Unit A4 Horn House Lane, Kirkby, Merseyside, L33 7YG
Merseyrider Ltd, Edge Lane, Liverpool, Merseyside, L7 9LL.
MTL(Heysham) Ltd, Edge Lane, Liverpool, Merseyside, L7 9LL.
MTL(Manchester) Ltd, Norton Street, Manchester, M40 8TR
Bolton Coachways & Travel Ltd, The Garage, 137a High Street, Bolton
Greater Manchester, BL3 6SR

Depots: Laird Street, Birkenhead; Cuthbert Street, Bolton; Middleton Road, Heysham; Norton Street, Miles Platting; Speke Road, Garston, Liverpool; East Lancashire Road, Gillmoss, Liverpool; Green Lane, Liverpool; Woodend Avenue, Speke, Liverpool; Jackson Street, St Helens and Canning Street, Southport.

0001	WCN643Y	Volvo B10M-56	Van Hool Alizée	C53F	1982	Ex Park's, Hamilton, 1990
0002	B483UNB	Volvo B10M-61	Van Hool Alizée	C53F	1985	Ex Shearings, 1991
0004	VWT134N	Bedford YRT	Duple Dominant	C53F	1974	Ex Hornby, Huyton, 1988
0015	VRG415T	MCW Metrobus DR101/4	MCW	H46/30F	1979	Ex London Buses, 1993
0016	VRG416T	MCW Metrobus DR101/4	MCW	H46/30F	1979	Ex London Buses, 1993
0017	VRG417T	MCW Metrobus DR101/4	MCW	H46/30F	1979	Ex London Buses, 1993
0018	VRG418T	MCW Metrobus DR101/4	MCW	H46/30F	1979	Ex London Buses, 1993
0019	UKA19V	MCW Metrobus DR101/3	MCW	H43/30F	1979	
0020	UKA20V	MCW Metrobus DR101/3	MCW	H43/30F	1980	
0021	UKA21V	MCW Metrobus DR103/2(Gardner)	MCW	H43/30F	1980	
0022	UKA22V	MCW Metrobus DR103/2(Gardner)	MCW	H43/30F	1980	
0023	UKA23V	MCW Metrobus DR103/2(Gardner)	MCW	H43/30F	1980	
0030	ACM704X	Leyland Olympian ONTL11/1R	Eastern Coach Works	H46/31F	1981	
0034	ACM708X	Leyland Olympian ONTL11/1R	Eastern Coach Works	H46/31F	1981	
0037	ACM709X	Leyland Olympian ONTL11/1R	Eastern Coach Works	H46/31F	1982	
0040	ACM712X	Leyland Olympian ONTL11/1R	Eastern Coach Works	H46/31F	1982	
0041	ACM772X	Leyland Olympian ONTL11/1R	Eastern Coach Works	H46/31F	1982	
0042	ACM770X	MCW Metrobus DR104/9(Gardner)	Alexander RL	H45/31F	1982	
0043	ACM771X	MCW Metrobus DR104/9(Gardner)	Alexander RL	H45/31F	1982	
0054	DEM821Y	Volvo-Ailsa B55-10	Alexander RV	H45/31F	1982	
0055	DEM822Y	Volvo-Ailsa B55-10	Alexander RV	H45/31F	1982	

0056-0063 MCW Metrobus DR102/29 Alexander RL H45/31F 1982

0056	EKA156Y	**0058**	DEM758Y	**0060**	DEM760Y	**0062**	DEM762Y	**0063**	DEM763Y
0057	EKA157Y	**0059**	DEM759Y	**0061**	DEM761Y				

0064-0068 Leyland Olympian ONTL11/1R Alexander RL DPH43/25F 1983

0064	A316GLV	**0065**	A317GLV	**0066**	A318GLV	**0067**	A319GLV	**0068**	A320GLV

0069-0081 Volvo B55-10 Alexander RV H44/37F 1984

0069	A151HLV	**0072**	A154HLV	**0075**	A157HLV	**0078**	A160HLV	**0080**	A162HLV
0070	A152HLV	**0073**	A155HLV	**0076**	A158HLV	**0079**	A161HLV	**0081**	A163HLV
0071	A153HLV	**0074**	A156HLV	**0077**	A159HLV				

0084	GHB84W	Bristol VRT/SL3/6LXB	East Lancashire	H44/32F	1981	Ex Blue Triangle, 1994
0093	H838GLD	Mercedes-Benz 609D	North West CS	B22F	1990	
0094	H839GLD	Mercedes-Benz 609D	North West CS	B22F	1990	
0095	G922EKF	Mercedes-Benz 609D	North West CS	B24F	1990	
0096	F373HMB	Mercedes-Benz 609D	North West CS	C19F	1989	
0097	F68GTU	Mercedes-Benz 609D	North West CS	C19F	1989	
0098	A863TCA	Mercedes-Benz L608D	Coachcraft	C21F	1983	Ex Ball, St Helens, 1989
0099	E685UVR	Mercedes-Benz 609D	North West CS	C19F	1988	
0108	OBN511R	Leyland Fleetline FE30AGR	Northern Counties	H43/32F	1976	Ex Cambus, 1990
0110	KSD110W	Volvo-Ailsa B55-10	Alexander AV	DPH40/31F	1980	Ex Blue Triangle, 1994
0112	KSD112W	Volvo-Ailsa B55-10	Alexander AV	DPH40/31F	1980	Ex Blue Triangle, 1994
0115	PRJ487R	Daimler Fleetline CRG6LXB	Northern Counties	H43/32F	1976	Ex Cambus, 1990
0116	WWH39L	Daimler Fleetline CRG6LXB	Park Royal	H43/32F	1973	Ex Greater Manchester, 1987
0119	OBN509R	Leyland Fleetline FE30AGR	Park Royal	H43/32F	1973	Ex Cambus, 1990

Showing fleet number 030 Leyland Olympian ACM704X is an Eastern Coach Works-bodied example now with the Fareway operation and liveried accordingly. While the computer lists show all fleet numbers with four digits, the application to vehicles depends on fleet number plates, transfers and operations. *Roy Marshall*

Cross-river services were introduced under the Silver Service name using buses fitted with high-back seating. Having emerged into Liverpool centre is 0067, A319GLV. *Phillip Stephenson*

120-146		Daimler Fleetline CRG6LXB		Northern Counties		H43/32F	1973-76	Ex Greater Manchester,1987 124/34 Ex Cambus, 1989	
121	VNB226L	**127**	YNA317M	**132**	YNA339M	**135**	YNA352M	**139**	YNA370M
122	VNB240L	**128**	YNA326M	**133**	YNA340M	**136**	YNA355M	**140**	GND490N
123	YNA271M	**129**	YNA333M	**134**	PRJ493R	**138**	YNA365M	**146**	BNE742N
124	PRJ499R	**131**	YNA336M						

149-155		Leyland Fleetline FE30AGR		MCW		H44/32F	1976/77	Ex Cumberland, 1988	
149	OJD194R	**151**	OJD197R	**153**	OJD170R	**155**	OJD169R	**157**	OJD224R
150	OJD203R	**152**	OJD245R	**154**	OUC39R				

159-168		Leyland Olympian ONCL10/2RZ		Northern Counties		H51/34F	1988		
159	F438AKB	**161**	F440AKB	**163**	F442AKB	**165**	F456BKF	**167**	F458BKF
160	F439AKB	**162**	F441AKB	**164**	F455BKF	**166**	F457BKF	**168**	F459BKF

169	PRJ495R	Daimler Fleetline CRG6LXB	Northern Counties	H43/32F	1976	Ex Cambus, 1990
170	OBN506R	Leyland Fleetline FE30AGR	Northern Counties	H43/32F	1976	Ex Cambus, 1990
172	E913KYR	Leyland Olympian ONLXB/1RH	Northern Counties	H43/30F	1988	Ex London Buses, 1991
173	E926KYR	Leyland Olympian ONLXB/1RH	Northern Counties	H43/30F	1988	Ex London Buses, 1991
174	E916KYR	Leyland Olympian ONLXB/1RH	Northern Counties	H43/30F	1988	Ex London Buses, 1991
175	E928KYR	Leyland Olympian ONLXB/1RH	Northern Counties	H43/30F	1988	Ex London Buses, 1991
176	OBN510R	Leyland Fleetline FE30AGR	Northern Counties	H43/32F	1976	Ex Cambus, 1990
177	OBN508R	Leyland Fleetline FE30AGR	Northern Counties	H43/32F	1976	Ex Cambus, 1990
178	JDB109N	Daimler Fleetline CRG6LXB	Northern Counties	H43/32F	1975	Ex Warrington, 1992
184	CFM84S	Leyland Fleetline FE30AGR	Northern Counties	H43/29F	1978	Ex Blue Triangle, 1994
190	LDS190A	Bristol VRT/SL3/501	Eastern Coach Works	DPH41/29F	1978	Ex Blue Triangle, 1994

201-215		Leyland Olympian ONCL10/1RZ		Northern Counties		H45/30F	1988		
201	E201WBG	**204**	E204WBG	**207**	E207WBG	**210**	E210WBG	**214**	E214WBG
202	E202WBG	**205**	E205WBG	**208**	E208WBG	**212**	E212WBG	**215**	E215WBG
203	E203WBG	**206**	E206WBG	**209**	E209WBG	**213**	E213WBG		

216-230		Leyland Olympian ONCL10/1RZ		Alexander RL		H45/30F	1988		
216	E216WBG	**219**	E219WBG	**222**	E222WBG	**225**	E225WBG	**228**	E228WBG
217	E217WBG	**220**	E220WBG	**223**	E223WBG	**226**	E226WBG	**229**	E229WBG
218	E218WBG	**221**	E221WBG	**224**	E224WBG	**227**	E227WBG	**230**	E230WBG

231-250		Leyland Olympian ONCL10/1RZ		Alexander RL		H45/30F	1989		
231	F231YTJ	**235**	F235YTJ	**239**	F239YTJ	**243**	F243YTJ	**247**	F247YTJ
232	F232YTJ	**236**	F236YTJ	**240**	F240YTJ	**244**	F244YTJ	**248**	F248YTJ
233	F233YTJ	**237**	F237YTJ	**241**	F241YTJ	**245**	F245YTJ	**249**	F249YTJ
234	F234YTJ	**238**	F238YTJ	**242**	F242YTJ	**246**	F246YTJ	**250**	F250YTJ

Merseybus took delivery of thirty Leyland Olympians in 1988 with bodywork divided between Northern Counties and Alexander. One of the Northern Counties examples is 215, E215WBG, seen in Fareway's livery as it passes Lime Street while heading for the city centre.
Keith Grimes

251-270	Leyland Olympian ONCL10/1RZ	Northern Counties	H47/30F	1989	

251	F251YTJ	**255**	F255YTJ	**259**	F259YTJ	**263**	F263YTJ	**267**	F267YTJ
252	F252YTJ	**256**	F256YTJ	**260**	F260YTJ	**264**	F264YTJ	**268**	F268YTJ
253	F253YTJ	**257**	F257YTJ	**261**	F261YTJ	**265**	F265YTJ	**269**	F269YTJ
254	F254YTJ	**258**	F258YTJ	**262**	F262YTJ	**266**	F266YTJ	**270**	F270YTJ

0300	ABN300L	Leyland Leopard PSU3B/4R	Plaxton Elite Express III	C53F	1973	Ex Bolton Coachways, 1994

368-405	Leyland Fleetline FE30AGR	MCW	H43/32F	1979	Ex Busways, 1993-94

368	VCU368T	**377**	VCU377T	**384**	VCU384T	**391**	VCU391T	**398**	VCU398T
369	VCU369T	**378**	VCU378T	**385**	VCU385T	**392**	VCU392T	**399**	VCU399T
370	VCU370T	**379**	VCU379T	**386**	VCU386T	**393**	VCU393T	**401**	VCU401T
371	VCU371T	**380**	VCU380T	**387**	VCU387T	**394**	VCU394T	**403**	VCU403T
374	VCU374T	**381**	VCU381T	**388**	VCU388T	**395**	VCU395T	**404**	VCU404T
375	VCU375T	**382**	VCU382T	**389**	VCU389T	**396**	VCU396T	**405**	VCU405T
376	VCU376T	**383**	VCU383T	**390**	VCU390T	**397**	VCU397T		

0526	SDA526S	Leyland Fleetline FE30AGR	MCW	H43/33F	1977	Ex Smalley, Heysham, 1993
0532	SDA632S	Leyland Fleetline FE30AGR	Park Royal	H43/33F	1977	Ex Smalley, Heysham, 1993
0533	SDA533S	Leyland Fleetline FE30AGR	MCW	H43/33F	1977	Ex Smalley, Heysham, 1993
0553	LMS153W	Leyland Fleetline FE30AGR	Alexander D	H44/31F	1980	Ex Smalley, Heysham, 1993
0554	NOC454R	Leyland Fleetline FE30AGR	MCW	H43/33F	1977	Ex Smalley, Heysham, 1993
0555	NOC455R	Leyland Fleetline FE30AGR	MCW	H43/33F	1977	Ex Smalley, Heysham, 1993
0596	NOC396R	Leyland Fleetline FE30ALR(6LXB)	MCW	H43/33F	1977	Ex Smalley, Heysham, 1993
0655	NMY655E	AEC Routemaster R2RH	Park Royal	H32/24F	1967	Ex Blue Triangle, 1994
0682	VWM82L	Leyland Atlantean AN68/1R	Alexander AL	O45/29D	1973	Ex Southport, 1974
0687	CWG687V	Leyland Atlantean AN68A/1R	Alexander AL	H45/29D	1973	Ex Blue Triangle, 1994
0742	NOC742R	Leyland Fleetline FE30AGR	East Lancashire	H43/33F	1977	Ex Smalley, Heysham, 1993

801-825	MCW Metrobus DR132	MCW	H46/31F	1989	

801	F801YLV	**806**	F806YLV	**811**	F811YLV	**816**	F816YLV	**821**	F821YLV
802	F802YLV	**807**	F807YLV	**812**	F812YLV	**817**	F817YLV	**822**	F822YLV
803	F803YLV	**808**	F808YLV	**813**	F813YLV	**818**	F818YLV	**823**	F823YLV
804	F804YLV	**809**	F809YLV	**814**	F814YLV	**819**	F819YLV	**824**	F824YLV
805	F805YLV	**810**	F810YLV	**815**	F815YLV	**820**	F820YLV	**825**	F825YLV

0851	XKC851K	Leyland Atlantean PDR2/1	Alexander AL	H51/31F	1971	Ex Blue Triangle, 1994
0858	XKC858K	Leyland Atlantean PDR2/1	Alexander AL	H51/31F	1971	Ex Blue Triangle, 1994
0861	XKC861K	Leyland Atlantean PDR2/1	Alexander AL	H51/31F	1971	Ex Blue Triangle, 1994

1001-1070	Leyland Atlantean AN68D/1R	Alexander AL	H43/32F	1983-84	

1001	A321GLV	**1015**	A335GLV	**1029**	A109HLV	**1043**	A123HLV	**1057**	A137HLV
1002	A322GLV	**1016**	A96HLV	**1030**	A110HLV	**1044**	A124HLV	**1058**	A138HLV
1003	A323GLV	**1017**	A97HLV	**1031**	A111HLV	**1045**	A125HLV	**1059**	A139HLV
1004	A324GLV	**1018**	A98HLV	**1032**	A112HLV	**1046**	A126HLV	**1060**	A140HLV
1005	A325GLV	**1019**	A99HLV	**1033**	A113HLV	**1047**	A127HLV	**1061**	A141HLV
1006	A326GLV	**1020**	A100HLV	**1034**	A114HLV	**1048**	A128HLV	**1062**	A142HLV
1007	A327GLV	**1021**	A101HLV	**1035**	A115HLV	**1049**	A129HLV	**1063**	A143HLV
1008	A328GLV	**1022**	A102HLV	**1036**	A116HLV	**1050**	A130HLV	**1064**	A144HLV
1009	A329GLV	**1023**	A103HLV	**1037**	A117HLV	**1051**	A131HLV	**1065**	A145HLV
1010	A330GLV	**1024**	A104HLV	**1038**	A118HLV	**1052**	A132HLV	**1066**	A146HLV
1011	A331GLV	**1025**	A105HLV	**1039**	A119HLV	**1053**	A133HLV	**1068**	A148HLV
1012	A332GLV	**1026**	A106HLV	**1040**	A120HLV	**1054**	A134HLV	**1069**	A149HLV
1013	A333GLV	**1027**	A107HLV	**1041**	A121HLV	**1055**	A135HLV	**1070**	B926KWM
1014	A334GLV	**1028**	A108HLV	**1042**	A122HLV	**1056**	A136HLV		

1105-1116	Leyland Atlantean AN68A/1R	East Lancashire	H43/32F	1976	Ex Blackburn, 1993

1105	PCW105P	**1108**	PCW108P	**1110**	PCW110P	**1112**	PCW112P	**1116**	PCW116P
1106	PCW106P	**1109**	PCW109P	**1111**	PCW111P	**1113**	PCW113P		

1118	THX618S	Leyland Fleetline FE30ALR	Park Royal	H44/27D	1978	Ex ??
1176	THX276S	Leyland Fleetline FE30ALR	MCW	H44/27D	1977	Ex ??

1203	KHF216P	Leyland Atlantean AN68A/1R	Northern Counties	H43/32F	1977	Ex Sheffield Omnibus, 1994
1204	KHF215P	Leyland Atlantean AN68A/1R	Northern Counties	H43/32F	1977	Ex Sheffield Omnibus, 1994
1214	KBU914P	Leyland Atlantean AN68A/1R	Northern Counties	H43/32F	1977	Ex Sheffield Omnibus, 1994
1215	LJA615P	Leyland Atlantean AN68A/1R	Northern Counties	H43/32F	1977	Ex Sheffield Omnibus, 1994
1219	LJA619P	Leyland Atlantean AN68A/1R	Northern Counties	H43/32F	1977	Ex Sheffield Omnibus, 1994
1236	LJA636P	Leyland Atlantean AN68A/1R	Northern Counties	H43/32F	1977	Ex Sheffield Omnibus, 1994
1263	BKC263K	Leyland Atlantean PDR1A/1	Alexander AL	H43/33F	1972	
1290	SJA790S	Leyland Atlantean AN68A/1R	Northern Counties	H43/32F	1977	Ex Sheffield Omnibus, 1994

1300-1393 Leyland Atlantean AN68/1R Alexander AL H43/32F 1972-73

1300	DKC300L	**1310**	DKC310L	**1329**	DKC329L	**1358**	DKC358L	**1369**	DKC369L
1301	DKC301L	**1311**	DKC311L	**1330**	DKC330L	**1359**	DKC359L	**1374**	DKC374L
1304	DKC304L	**1313**	DKC313L	**1331**	DKC331L	**1360**	DKC360L	**1378**	DKC378L
1305	DKC305L	**1314**	DKC314L	**1340**	DKC340L	**1361**	DKC361L	**1382**	DKC382L
1306	DKC306L	**1315**	DKC315L	**1342**	DKC342L	**1364**	DKC364L	**1391**	DKC391L
1307	DKC307L	**1318**	DKC318L	**1349**	DKC349L	**1365**	DKC365L	**1392**	DKC392L
1308	DKC308L	**1325**	DKC325L	**1353**	DKC353L	**1368**	DKC368L	**1393**	DKC393L
1309	DKC309L	**1326**	DKC326L	**1356**	DKC356L				

1447-1544 Leyland Atlantean AN68/1R Alexander AL H43/32F* 1973-74 *1449/1524 are O43/32F

1447	GKA447L	**1466**	GKA466L	**1482**	GKA482L	**1501**	GKA501M	**1517**	GKA517M
1448	GKA448L	**1467**	GKA467L	**1483**	GKA483L	**1503**	GKA503M	**1518**	GKA518M
1449	GKA449L	**1468**	GKA468L	**1484**	GKA484L	**1504**	GKA504M	**1522**	GKA522M
1451	GKA451L	**1469**	GKA469L	**1486**	GKA486M	**1507**	GKA507M	**1524**	GKA524M
1454	GKA454L	**1470**	GKA470L	**1488**	GKA488M	**1508**	GKA508M	**1529**	GKA529M
1456	GKA456L	**1471**	GKA471L	**1489**	GKA489M	**1509**	GKA509M	**1531**	GKA531M
1457	GKA457L	**1472**	GKA472L	**1490**	GKA490M	**1510**	GKA510M	**1535**	GKA535M
1459	GKA459L	**1473**	GKA473L	**1491**	GKA491M	**1511**	GKA511M	**1536**	GKA536M
1460	GKA460L	**1474**	GKA474L	**1492**	GKA492M	**1512**	GKA512M	**1537**	GKA537M
1461	GKA461L	**1476**	GKA476L	**1493**	GKA493M	**1513**	GKA513M	**1538**	GKA539M
1462	GKA462L	**1477**	GKA477L	**1495**	GKA495M	**1514**	GKA514M	**1540**	GKA540M
1463	GKA463L	**1479**	GKA479L	**1498**	GKA498M	**1515**	GKA515M	**1541**	GKA541M
1464	GKA464L	**1480**	GKA480L	**1499**	GKA499M	**1516**	GKA516M	**1544**	GKA544M
1465	GKA465L								

1551-1585 Leyland Atlantean AN68/1R Alexander AL H43/32F* 1974 *1551 is O43/32F

1551	OLV551M	**1555**	PKB555M	**1560**	PKB560M	**1567**	RKB567M	**1574**	RKA574N
1553	OLV553M	**1558**	PKB558M	**1563**	PKB563M	**1570**	RKA570N	**1585**	RKA589N
1554	PKB554M								

A happy load of holiday makers take advantage of the fine weather on Southport's open-top service. Turning off the fashionable Lord Street is 1612, GKA37N, now sporting the latest livery for Southport that gives more prominence to the town's name and former livery.
Steve Sanderson

1587-1624 Leyland Atlantean AN68/1R Alexander AL H43/32F* 1974 *1612 is O43/32F

1587	GKA12N	**1592**	GKA17N	**1597**	GKA22N	**1602**	GKA27N	**1615**	GKA40N
1588	GKA13N	**1593**	GKA18N	**1598**	GKA23N	**1603**	GKA28N	**1618**	GKA43N
1589	GKA14N	**1594**	GKA19N	**1599**	GKA24N	**1610**	GKA35N	**1619**	GKA44N
1590	GKA15N	**1595**	GKA20N	**1600**	GKA25N	**1612**	GKA37N	**1622**	GKA47N
1591	GKA16N	**1596**	GKA21N	**1601**	GKA26N	**1613**	GKA38N	**1624**	GKA49N

1626-1685 Leyland Atlantean AN68/1R Alexander AL H43/32F 1976

1626	HTJ626P	**1634**	HTJ634P	**1642**	HTJ642P	**1655**	HTJ655P	**1675**	HTJ675P
1627	HTJ627P	**1635**	HTJ635P	**1643**	HTJ643P	**1658**	HTJ658P	**1676**	HTJ676P
1628	HTJ628P	**1636**	HTJ636P	**1644**	HTJ644P	**1659**	HTJ659P	**1682**	HTJ682P
1630	HTJ630P	**1637**	HTJ637P	**1646**	HTJ646P	**1664**	HTJ664P	**1683**	HTJ683P
1631	HTJ631P	**1638**	HTJ638P	**1649**	HTJ649P	**1669**	HTJ669P	**1684**	HTJ684P
1632	HTJ632P	**1639**	HTJ639P	**1653**	HTJ653P	**1670**	HTJ670P	**1685**	HTJ685P
1633	HTJ633P	**1640**	HTJ640P	**1654**	HTJ654P	**1674**	HTJ674P		

1687-1725 Leyland Atlantean AN68/1R East Lancashire H43/32F 1976

1687	JWM687P	**1697**	JWM697P	**1704**	JWM704P	**1712**	JWM712P	**1721**	JWM721P
1689	JWM689P	**1698**	JWM698P	**1705**	JWM705P	**1713**	JWM713P	**1722**	JWM722P
1693	JWM693P	**1700**	JWM700P	**1706**	JWM706P	**1714**	JWM714P	**1723**	JWM723P
1694	JWM694P	**1701**	JWM701P	**1708**	JWM708P	**1715**	JWM715P	**1724**	JWM724P
1695	JWM695P	**1702**	JWM702P	**1709**	JWM709P	**1717**	JWM717P	**1725**	JWM725P
1696	JWM696P	**1703**	JWM703P	**1710**	JWM710P	**1720**	JWM720P		

1726-1765 Leyland Atlantean AN68A/1R East Lancashire H43/32F 1976-77

1726	LKF726R	**1736**	LKF736R	**1745**	LKF745R	**1752**	LKF752R	**1759**	MTJ759S
1727	LKF727R	**1737**	LKF737R	**1746**	LKF746R	**1753**	LKF753R	**1760**	MTJ760S
1728	LKF728R	**1738**	LKF738R	**1747**	LKF747R	**1754**	LKF754R	**1761**	MTJ761S
1730	LKF730R	**1740**	LKF740R	**1748**	LKF748R	**1755**	LKF755R	**1762**	MTJ762S
1731	LKF731R	**1741**	LKF741R	**1749**	LKF749R	**1756**	MTJ756S	**1763**	MTJ763S
1732	LKF732R	**1742**	LKF742R	**1750**	LKF750R	**1757**	MTJ757S	**1764**	MTJ764S
1734	LKF734R	**1743**	LKF743R	**1751**	LKF751R	**1758**	MTJ758S	**1765**	MTJ765S
1735	LKF735R	**1744**	LKF744R						

1777-1813 Leyland Atlantean AN68/1R MCW H43/32F 1978

1777	OEM777S	**1781**	OEM781S	**1795**	OEM795S	**1803**	OEM803S	**1807**	PKA721S
1778	OEM778S	**1786**	OEM786S	**1796**	OEM796S	**1805**	OEM805S	**1808**	PKA722S
1779	OEM779S	**1791**	OEM791S	**1800**	OEM800S	**1806**	PKA720S	**1813**	PKA727S
1780	OEM780S								

The latest version of the Mereyrider livery is seen on East Lancashire-bodied Leyland Atlantean. As shown at the end of the fleet list, many of this type are now in the Merseyrider and MTL Manchester operations.
Phillip Stephenson

1821-1842 Leyland Atlantean AN68A/1R East Lancashire H45/33F* 1979-80 *1840 is H43/32F

1821	RBG821T	**1826**	RBG826T	**1830**	TWM214V	**1834**	WWM917W	**1839**	RBG839T
1822	TWM209V	**1827**	RBG827T	**1831**	TWM215V	**1835**	TWM219V	**1840**	RBG840T
1823	TWM210V	**1828**	TWM212V	**1832**	TWM216V	**1836**	TWM220V	**1841**	RBG841T
1824	TWM211V	**1829**	TWM213V	**1833**	TWM217V	**1837**	RBG837T	**1842**	TWM221V
1825	RBG825T								

1847	WWM924W	Leyland Atlantean AN68B/1R	Willowbrook	DPH45/39F	1980
1867	AFY187X	Leyland Atlantean AN68B/1R	Willowbrook	H45/33F	1982

1873-1967 Leyland Atlantean AN68B/1R Alexander AL H43/32F 1981-82

1873	XEM873W	**1892**	XEM892W	**1911**	XEM911W	**1930**	ACM730X	**1949**	ACM749X
1874	XEM874W	**1893**	XEM893W	**1912**	XEM912W	**1931**	ACM731X	**1950**	ACM750X
1875	XEM875W	**1894**	XEM894W	**1913**	ACM713X	**1932**	ACM732X	**1951**	ACM751X
1876	XEM876W	**1895**	XEM895W	**1914**	ACM714X	**1933**	ACM733X	**1952**	ACM752X
1877	XEM877W	**1896**	XEM896W	**1915**	ACM715X	**1934**	ACM734X	**1953**	ACM753X
1878	XEM878W	**1897**	XEM897W	**1916**	ACM716X	**1935**	ACM735X	**1954**	ACM754X
1879	XEM879W	**1898**	XEM898W	**1917**	ACM717X	**1936**	ACM736X	**1955**	ACM755X
1880	XEM880W	**1899**	XEM899W	**1918**	ACM718X	**1937**	ACM737X	**1956**	ACM756X
1881	XEM881W	**1900**	XEM900W	**1919**	ACM719X	**1938**	ACM738X	**1957**	ACM757X
1882	XEM882W	**1901**	XEM901W	**1920**	ACM720X	**1939**	ACM739X	**1958**	ACM758X
1883	XEM883W	**1902**	XEM902W	**1921**	ACM721X	**1940**	ACM740X	**1959**	ACM759X
1884	XEM884W	**1903**	XEM903W	**1922**	ACM722X	**1941**	ACM741X	**1960**	ACM760X
1885	XEM885W	**1904**	XEM904W	**1923**	ACM723X	**1942**	ACM742X	**1961**	ACM761X
1886	XEM886W	**1905**	XEM905W	**1924**	ACM724X	**1943**	ACM743X	**1962**	ACM762X
1887	XEM887W	**1906**	XEM906W	**1925**	ACM725X	**1944**	ACM744X	**1963**	ACM763X
1888	XEM888W	**1907**	XEM907W	**1926**	ACM726X	**1945**	ACM745X	**1964**	ACM764X
1889	XEM889W	**1908**	XEM908W	**1927**	ACM727X	**1946**	ACM746X	**1965**	ACM765X
1890	XEM890W	**1909**	XEM909W	**1928**	ACM728X	**1947**	ACM747X	**1966**	ACM766X
1891	XEM891W	**1910**	XEM910W	**1929**	ACM729X	**1948**	ACM748X	**1967**	ACM767X

1968-1991 Leyland Atlantean AN68D/1R Alexander AL H43/32F 1982

1968	ACM768X	**1974**	DEM774Y	**1979**	DEM779Y	**1984**	DEM784Y	**1988**	DEM788Y
1969	ACM769X	**1975**	DEM775Y	**1980**	DEM780Y	**1985**	DEM785Y	**1989**	DEM789Y
1970	CKB170X	**1976**	DEM776Y	**1981**	DEM781Y	**1986**	DEM786Y	**1990**	DEM790Y
1971	EKA171Y	**1977**	DEM777Y	**1982**	DEM782Y	**1987**	DEM787Y	**1991**	DEM791Y
1973	DEM773Y	**1978**	DEM778Y	**1983**	DEM783Y				

2052-2249 Leyland Titan TNLXB-2RRSp Park Royal H44/30F* 1979-80 Ex London Buses, 1992-94
* being changed ex H44/24D

2052	WYV52T	**2123**	CUL123V	**2157**	CUL157V	**2196**	CUL196V	**2219**	CUL219V
2059	WYV59T	**2124**	CUL124V	**2159**	CUL159V	**2199**	CUL199V	**2220**	CUL220V
2065	WYV65T	**2125**	CUL125V	**2166**	CUL166V	**2200**	CUL200V	**2231**	EYE231V
2070	CUL70V	**2127**	CUL127V	**2171**	CUL171V	**2201**	CUL201V	**2232**	EYE232V
2073	CUL73V	**2132**	CUL132V	**2174**	CUL174V	**2203**	CUL203V	**2234**	EYE234V
2104	CUL104V	**2136**	CUL136V	**2177**	CUL177V	**2204**	CUL204V	**2239**	EYE239V
2107	CUL107V	**2144**	CUL144V	**2184**	CUL184V	**2211**	CUL211V	**2241**	EYE241V
2108	CUL108V	**2145**	CUL145V	**2187**	CUL187V	**2213**	CUL213V	**2245**	EYE245V
2109	CUL109V	**2147**	CUL147V	**2188**	CUL188V	**2217**	CUL217V	**2247**	EYE247V
2115	CUL115V	**2148**	CUL148V	**2192**	CUL192V	**2218**	CUL218V	**2249**	EYE249V
2118	CUL118V	**2155**	CUL155V	**2194**	CUL194V				

2253-2278 Leyland Titan TNLXB-2RRSp PRV/Leyland H44/30F* 1981 Ex London Buses, 1992-93
* being changed ex H44/24D

2253	GYE253W	**2256**	GYE256W	**2258**	GYE258W	**2269**	GYE269W	**2278**	GYE278W
2255	GYE255W	**2257**	GYE257W	**2259**	GYE259W	**2276**	GYE276W		

Leyland Titans continue to arrive, with many examples passing through the works programme to convert them to single door. Early repaints were in the maroon livery, but more recently vehicles are being outshopped from Edge Lane works in the latest scheme. Seen heading for the Netherley area of the city is 2194, CUL 194V. *Phillip Stephenson.*

2284-2547 Leyland Titan TNLXB-2RRSp Leyland H44/30F* 1981-82 Ex London Buses, 1992-94
* being changed ex H44/24D

2284	KYN284X	**2339**	KYV339X	**2376**	KYV376X	**2416**	KYV416X	**2464**	KYV464X
2289	KYN289X	**2344**	KYV344X	**2382**	KYV382X	**2417**	KYV417X	**2468**	KYV468X
2293	KYN293X	**2347**	KYV347X	**2383**	KYV383X	**2418**	KYV418X	**2472**	KYV472X
2299	KYN299X	**2350**	KYV350X	**2385**	KYV385X	**2421**	KYV421X	**2483**	KYV483X
2301	KYN301X	**2351**	KYV351X	**2389**	KYV389X	**2424**	KYV424X	**2484**	KYV484X
2303	KYN303X	**2353**	KYV353X	**2391**	KYV391X	**2425**	KYV425X	**2489**	KYV489X
2304	KYN304X	**2354**	KYV354X	**2398**	KYV398X	**2427**	KYV427X	**2491**	KYV491X
2316	KYN316X	**2355**	KYV355X	**2400**	KYV400X	**2430**	KYV430X	**2494**	KYV494X
2324	KYV324X	**2359**	KYV359X	**2402**	KYV402X	**2431**	KYV431X	**2499**	KYV499X
2330	KYV330X	**2363**	KYV363X	**2407**	KYV407X	**2440**	KYV440X	**2509**	KYV509X
2332	KYV332X	**2364**	KYV364X	**2409**	KYV409X	**2443**	KYV443X	**2528**	KYV528X
2335	JGH118X	**2365**	KYV365X	**2412**	KYV412X	**2449**	KYV449X	**2534**	KYV534X
2337	KYV337X	**2367**	KYV367X	**2413**	KYV413X	**2450**	KYV450X	**2547**	KYV547X
2338	KYV338X	**2374**	KYV374X	**2414**	KYV414X	**2463**	KYV463X		

2561-2884 Leyland Titan TNLXB-2RR* Leyland H44/30F* 1982-84 Ex London Buses, 1992-94
*2878/9/84 are TNTL112RR, * being changed ex H44/24D

2561	NUW561Y	**2628**	NUW628Y	**2682**	OHV682Y	**2726**	OHV726Y	**2754**	OHV754Y
2570	NUW570Y	**2638**	NUW638Y	**2690**	OHV690Y	**2730**	OHV730Y	**2878**	A878SUL
2599	NUW599Y	**2655**	NUW655Y	**2698**	OHV698Y	**2741**	OHV741Y	**2879**	A879SUL
2612	NUW612Y	**2656**	NUW656Y	**2703**	OHV703Y	**2746**	OHV746Y	**2884**	A884SUL

3003	GBU3V	MCW Metrobus DR101/6	MCW	H43/30F	1979	Ex Stevensons, 1994
3010	GBU10V	MCW Metrobus DR101/6	MCW	H43/30F	1979	Ex Stevensons, 1993
3016	WYW16T	MCW Metrobus DR101/8	MCW	H43/28D	1979	Ex London Buses, 1993
3026	WYW26T	MCW Metrobus DR101/8	MCW	H43/28D	1979	Ex London Buses, 1993
3037	WYW37T	MCW Metrobus DR101/8	MCW	H43/28D	1979	Ex London Buses, 1993
3039	A639THV	Leyland Titan TNLXB-2RR	Leyland	H44/24D	1983	Ex London Buses, 1993

3074-3081		MCW Metrobus DR102/20		MCW		H46/31F	1981	Ex Newport, 1994

3074	JBO74W	**3077**	JBO77W	**3078**	JBO78W	**3079**	JBO79W	**3081**	JBO81W
3076	JBO76W								

3106	BYX106V	MCW Metrobus DR101/9	MCW	H43/28D	1979	Ex London Buses, 1993
3108	BYX108V	MCW Metrobus DR101/9	MCW	H43/28D	1979	Ex London Buses, 1993
3138	JHE138W	MCW Metrobus DR104/6	MCW	H46/31F	1981	Ex Stevensons, 1994
3144	JHE144W	MCW Metrobus DR104/6	MCW	H46/31F	1981	Ex Stevensons, 1993
3152	JHE152W	MCW Metrobus DR104/6	MCW	H46/31F	1981	Ex Stevensons, 1994
3156	JHE156W	MCW Metrobus DR104/6	MCW	H46/31F	1981	Ex Stevensons, 1994
3229	BYX229V	MCW Metrobus DR101/12	MCW	H43/28D	1980	Ex London Buses, 1993
3312	BYX312V	MCW Metrobus DR101/12	MCW	H43/28D	1980	Ex London Buses, 1993
3377	GYE377W	MCW Metrobus DR101/12	MCW	H43/28D	1980	Ex London Buses, 1993
3501	GYE501W	MCW Metrobus DR101/12	MCW	H43/28D	1980	Ex London Buses, 1993
5202	AHH202T	Leyland National 10351B/1R		B44F	1978	Ex Cumberland, 1993
5203	AHH203T	Leyland National 10351B/1R		B44F	1978	Ex Cumberland, 1993
5204	TRN804V	Leyland National 10351B/1R		B44F	1979	Ex Cumberland, 1993
5205	AHH205T	Leyland National 10351B/1R		B44F	1978	Ex Cumberland, 1993
5208	YBO148T	Leyland National 10351A/1R		B44F	1979	Ex Rhondda, 1994
5209	SPC269R	Leyland National 10351A/1R		B41F	1977	Ex Luton & District, 1994
5210	AKY610T	Leyland National 10351B/1R		B41F	1979	Ex East Midland, 1994
5211	AKY611T	Leyland National 10351B/1R		B41F	1979	Ex East Midland, 1994
5212	AKY612T	Leyland National 10351B/1R		B41F	1979	Ex East Midland, 1994
5213	CHH213T	Leyland National 10351B/1R		B44F	1978	Ex Cumberland, 1993
5223	UPB323S	Leyland National 10351A/1R		B41F	1977	Ex Luton & District, 1994
5227	UTX727S	Leyland National 10351A/1R		B41F	1978	Ex Red & White, 1994
5230	DDW430V	Leyland National 10351A/1R		B44F	1979	Ex Red & White, 1994
5261	PJT261R	Leyland National 10351A/1R		B41F	1976	Ex BHT Buses, Bournemouth, 1994
5263	LDW363P	Leyland National 10351/1R		B41F	1975	Ex Red & White, 1994
5287	BYW387V	Leyland National 10351A/2R		B44F	1979	Ex Blue Triangle, 1994

6001-6012		Leyland National 11351A/1R	B49F	1979	6001/11 are fitted with Volvo engines

6001	SKF1T	**6005**	SKF5T	**6007**	SKF7T	**6009**	SKF9T	**6011**	SKF11T
6002	SKF2T	**6006**	SKF6T	**6008**	SKF8T	**6010**	SKF10T	**6012**	SKF12T
6003	SKF3T								

6013	DAK613V	Leyland National 11351A/1R	B49F	1979	Ex East Midland, 1994
6014	SKF14T	Leyland National 11351A/1R	B49F	1979	
6015	SKF15T	Leyland National 11351A/1R	B49F	1979	
6016	OOX816R	Leyland National 11351A/1R	DP45F	1979	Ex West Midlands Travel, 1994
6017	SKF17T	Leyland National 11351A/1R	B49F	1979	
6019	SKF19T	Leyland National 11351A/1R(Volvo)	B49F	1979	
6020	YFB970V	Leyland National 11351A/1R	B52F	1979	Ex Badgerline, 1994
6021	YFB971V	Leyland National 11351A/1R	B52F	1979	Ex Badgerline, 1994
6023	KHT123P	Leyland National 11351/1R	B52F	1976	Ex Badgerline, 1994

Vehicles based in the Lancaster area carry Haysham Travel names on a blue livery. Photographed in Lancaster is 6062, HCW762N, a Leyland National, one of a pair new to Fishwick in 1975. *Ken Grime.*

Fareway's livery is carried by 7001, RKA886T. Once fitted with high-back seating, the vehicle now has standard bus seating. It is seen heading north out of Liverpool towards Kirkby.
Phillip Stephenson

6025	FBV525S	Leyland National 11351A/1R	B49F	1978	Ex Fishwick, 1993
6027	CBV797S	Leyland National 11351A/1R	B49F	1977	Ex Ribble, 1994
6028	GLP428T	Leyland National 11351A/3R	B49F	1979	Ex Capital Citybus, 1993
6032	UHG732R	Leyland National 11351A/1R	B49F	1976	Ex Ribble, 1994
6033	TTC533T	Leyland National 11351A/1R	B52F	1978	Ex Badgerline, 1994

6035-6041 Leyland National 2 NL116L11/1R B49F 1979-80

6035	UEM35V	**6037**	UEM37V	**6039**	UEM39V	**6040**	UEM40V	**6041**	UEM41V
6036	UEM36V	**6038**	UEM38V						

6042	CFM342S	Leyland National 11351A/1R(6HLXB)	B49F	1978	Ex Smalley, Heysham, 1993
6044	URA604S	Leyland National 11351A/1R	B49F	1977	Ex East Midland, 1994
6047	BAL607T	Leyland National 11351A/1R	B49F	1978	Ex East Midland, 1994
6048	BAL608T	Leyland National 11351A/1R	B49F	1978	Ex East Midland, 1994
6049	BAL609T	Leyland National 11351A/1R	B49F	1978	Ex East Midland, 1994
6051	UTU981R	Leyland National 11351A/1R	B49F	1977	Ex East Midland, 1994
6054	PVF354R	Leyland National 11351A/1R	B52F	1977	Ex BHT Buses, Bournemouth, 1994
6055	JHW105P	Leyland National 11351A/1R	B52F	1975	Ex Badgerline, 1994
6056	XCW956R	Leyland National 11351A/1R	B49F	1977	Ex Fishwick, 1993
6057	KHT117P	Leyland National 11351A/1R	B52F	1975	Ex Badgerline, 1994
6061	WUH161T	Leyland National 11351A/1R	B49F	1978	Ex Red & White, 1994
6062	HCW762N	Leyland National 11351/1R	B49F	1975	Ex Raja, Blackburn, 1993
6064	NWO464R	Leyland National 11351A/1R	DP48F	1977	Ex Red & White, 1994
6066	PVF366R	Leyland National 11351A/1R	B52F	1976	Ex BHT Buses, Bournemouth, 1994
6067	VKE567S	Leyland National 11351A/1R	B49F	1977	Ex East Midland, 1994
6069	PJT269R	Leyland National 11351A/1R	B49F	1976	Ex Armstrong, Castle Douglas, 1993
6070	VKE570S	Leyland National 11351A/1R	B49F	1977	Ex East Midland, 1994
6071	HCW761N	Leyland National 11351/1R	B49F	1975	Ex Fishwick, 1993
6073	PVF373R	Leyland National 11351A/1R	B52F	1977	Ex BHT Buses, Bournemouth, 1994
6074	NWO474R	Leyland National 11351A/1R	B49F	1976	Ex Red & White, 1994
6075	PCD75R	Leyland National 11351A/1R	B49F	1977	Ex Victoria Travel, Earlstown, 1994
6077	BSF767S	Leyland National 11351A/1R	B49F	1978	Ex Burrows, Ogmore Vale, 1993

6080-6086 Leyland National 2 NL116L11/1R B49F 1980

6080	VBG80V	**6082**	VBG82V	**6083**	VBG83V	**6085**	VBG85V	**6086**	VBG86V
6081	VBG81V								

6087	VFX987S	Leyland National 11351A/1R	B49F	1978	Ex Burrows, Ogmore Vale, 1993
6088	UHG748R	Leyland National 11351A/1R	B49F	1976	Ex Cumberland, 1993
6091	RAU601R	Leyland National 11351A/1R	B49F	1976	Ex East Midland, 1994
6092	OTO592R	Leyland National 11351A/1R	B49F	1976	Ex East Midland, 1994
6093	PCK193P	Leyland National 11351A/1R	B49F	1976	Ex Fishwick, 1993

6095-6138		Leyland National 2 NL116L11/1R				B49F	1980		
6095	VBG95V	**6105**	VBG105V	**6114**	VBG114V	**6122**	VBG122V	**6131**	WWM910W
6097	VBG97V	**6106**	VBG106V	**6115**	VBG115V	**6124**	WWM907W	**6132**	WWM911W
6098	VBG98V	**6107**	VBG107V	**6116**	VBG116V	**6125**	WWM908W	**6133**	VBG133V
6099	VBG99V	**6108**	VBG108V	**6117**	VBG117V	**6126**	WWM909W	**6134**	WWM912W
6100	WWM905W	**6109**	VBG109V	**6118**	VBG118V	**6127**	VBG127V	**6135**	WWM913W
6101	VBG101V	**6110**	VBG110V	**6119**	VBG119V	**6128**	VBG128V	**6136**	WWM914W
6102	VBG102V	**6111**	VBG111V	**6120**	VBG120V	**6129**	VBG129V	**6137**	WWM915W
6103	VBG103V	**6112**	VBG112V	**6121**	VBG121V	**6130**	VBG130V	**6138**	WWM916W
6104	VBG104V	**6113**	VBG113V						

6139-6172		Leyland National 2 NL116AL11/1R				B49F*	1981-82 *6147-6151 are B23DL		
6139	XLV139W	**6146**	XLV146W	**6153**	XLV153W	**6160**	XLV160W	**6167**	CKB164X
6140	XLV140W	**6147**	XLV147W	**6154**	XLV154W	**6161**	XLV161W	**6168**	CKB165X
6141	XLV141W	**6148**	XLV148W	**6155**	XLV155W	**6162**	XLV162W	**6169**	CKB166X
6142	XLV142W	**6149**	XLV149W	**6156**	XLV156W	**6163**	XLV163W	**6170**	CKB167X
6143	XLV143W	**6150**	XLV150W	**6157**	XLV157W	**6164**	CKB161X	**6171**	CKB168X
6144	XLV144W	**6151**	XLV151W	**6158**	XLV158W	**6165**	CKB162X	**6172**	CKB169X
6145	XLV145W	**6152**	XLV152W	**6159**	XLV159W	**6166**	CKB163X		

6177	RSG817V	Leyland National 2 NL116L11/1R(6HLXB)		B52F	1980	Ex Fife Scottish, 1993
6178	TTC538T	Leyland National 11351A/1R		B52F	1979	Ex Badgerline, 1994
6179	TTC539T	Leyland National 11351A/1R		B52F	1979	Ex Badgerline, 1994
6180	TTC540T	Leyland National 11351A/1R		B52F	1979	Ex Badgerline, 1994
6185	AAS185W	Leyland National 2 NL116L11/1R		B48D	1980	Ex Yorkshire Rider, 1994
6186	WGY586S	Leyland National 11351A/3R		B49F	1978	Ex Capital Citybus, 1993
6187	WGY587S	Leyland National 11351A/3R		B49F	1978	Ex Capital Citybus, 1993
6190	OHW490R	Leyland National 11351A/1R		B52F	1977	Ex Badgerline, 1994
6192	OHW492R	Leyland National 11351A/1R		B52F	1977	Ex Badgerline, 1994
6197	NFB597R	Leyland National 11351A/1R		B52F	1976	Ex Badgerline, 1994
6200	MTJ773S	Leyland National 11351A/1R		B49F	1978	
6211	LUA311V	Leyland National 2 NL106L11/1R		B41F	1980	Ex Blue Triangle, 1994
6301	L301TEM	Volvo B10B	Alexander Strider	B49F	1994	
6302	L302TEM	Volvo B10B	Alexander Strider	B49F	1994	
6303	L303TEM	Volvo B10B	Alexander Strider	B49F	1994	

6330-6334		DAF SB220LC550		Ikarus CitiBus		B48F	1993		
6330	K130TCP	**6331**	K131TCP	**6332**	K132TCP	**6333**	K133TCP	**6334**	K510RJX

MTL have made a significant investment in new buses since the last edition of this book, with major developments in low-floor operation, and improved passenger comfort. Merseybus commenced operation of twelve Neoplan N4016 low-floor single deck buses in February operating them in Smart livery. 6407, L407TKB, *(opposite, top)* **was photographed in Liverpool. Also new in 1994 were fifty Volvo B6s with Plaxton Pointer bodywork. These are allocated across the operating companies. 7239, L239TKA,** *(opposite.bottom)* **is seen in Wirral Peninsula livery.** *Phillip Stephenson*

Photographed turning into Derby Square, Liverpool 6302, L302TEM, is one of three Volvo B10Bs with Alexander Strider bodywork. It is seen wearing Merseyrider livery.

003
City Circular
Anti-Clockwise
Wallasey (Tunnel)
Manchester (M62)
North & East
NEOPLAN
SMART
Merseytravel
L407 TKB

BEECHWOOD CIRCULAR
98
Wirral Peninsula
Merseybus
L239 TKA
PLAXTON

6402-6413	Aüwaerter Neoplan N4016		B33F	1993-94

6402	L402TKB	**6405**	L405TKB	**6408**	L408TKB	**6410**	L410TKB	**6412**	L412UFY
6403	L403TKB	**6406**	L406TKB	**6409**	L409TKB	**6411**	L411UFY	**6413**	L413TKB
6404	L404TKB	**6407**	L407TKB						

6501-6543	Volvo B10B-58	Wright Endurance	DP49F	1994

6501	L501TKA	**6510**	L510TKA	**6519**	L519TKA	**6528**	L528TKA	**6536**	L536TKA
6502	L502TKA	**6511**	L511TKA	**6520**	L520TKA	**6529**	L529TKA	**6537**	L537TKA
6503	L503TKA	**6512**	L512TKA	**6521**	L521TKA	**6530**	L530TKA	**6538**	L538TKA
6504	L504TKA	**6513**	L513TKA	**6522**	L522TKA	**6531**	L531TKA	**6539**	L539TKA
6505	L505TKA	**6514**	L514TKA	**6523**	L523TKA	**6532**	L532TKA	**6540**	L540TKA
6506	L506TKA	**6515**	L515TKA	**6524**	L524TKA	**6533**	L533TKA	**6541**	L541TKA
6507	L507TKA	**6516**	L516TKA	**6525**	L525TKA	**6534**	L534TKA	**6542**	L542TKA
6508	L508TKA	**6517**	L517TKA	**6526**	L526TKA	**6535**	L535TKA	**6543**	L543TKA
6509	L509TKA	**6518**	L518TKA	**6527**	L527TKA				

6908	COV8V	Leyland Leopard PSU3E/4R	Willowbrook 003	DP53F	1978	Ex Burrows, Ogmore Vale, 1993
6911	WRN11R	Leyland Leopard PSU4D/4R	Alexander AY	DP45F	1977	Ex Smalley, Heysham, 1993
6911	WRN12R	Leyland Leopard PSU4D/4R	Alexander AY	DP45F	1977	Ex Smalley, Heysham, 1993
6927	GSG127T	Leyland Leopard PSU3E/4R	Duple Dominant I	C49F	1978	Ex Fife Scottish, 1993
6930	GSG130T	Leyland Leopard PSU3E/4R	Duple Dominant I	C49F	1978	Ex Fife Scottish, 1993
6939	UDW639Y	Leyland Leopard PSU5E/4R	Eastern Coach Works	C53F	1982	Ex Blue Triangle, 1994
6946	GBO246W	Leyland Leopard PSU4E/4R	Duple Dominant	B47F	1981	Ex Smalley, Heysham, 1993
6963	AAX563A	Leyland Leopard PSU3E/4R	Eastern Coach Works	C53F	1982	Ex Blue Triangle, 1994
6969	WWY69X	Leyland Leopard PSU3E/4R	Willowbrook 003	C49F	1982	Ex Smalley, Heysham, 1993
6995	RJT146R	Leyland National 11351A/1R		B49F	1979	Ex Stagecoach South, 1993
6996	KDW354P	Leyland National 11351/1R/SC		DP48F	1975	Ex Red & White, 1994
6997	LUP897T	Leyland National 11351A/1R		B49F	1979	Ex Tees & District, 1993
6998	TPE170S	Leyland National 11351A/1R		B49F	1978	Ex Wycombe Bus Company, 1993
6999	GSX864T	Leyland National 11351A/1R		B49F	1979	Ex Burrows, Ogmore Vale, 1993
7000	OHF858S	Leyland National 11351A/1R		DP45F	1978	
7001	RKA886T	Leyland National 11351A/1R		B49F	1978	
7002	RKA887T	Leyland National 11351A/1R		DP45F	1978	
7004	XTJ4W	Leyland National 2 NL116AL11/1R		DP45F	1981	
7005	XTJ5W	Leyland National 2 NL116AL11/1R		B49F	1981	
7006	XTJ6W	Leyland National 2 NL116AL11/1R		DP45F	1981	
7007	XTJ7W	Leyland National 2 NL116AL11/1R		B49F	1981	

7009-7020	Leyland Tiger TRCTL11/2R	Duple Dominant IV Express	C49F	1982

7009	CKC623X	**7012**	CKC626X	**7014**	CKC628X	**7017**	EKA217Y	**7019**	EKA219Y
7010	VKB708	**7013**	CKC624X	**7015**	EKA215Y	**7018**	EKA218Y	**7020**	EKA220Y
7011	CKC627X								

7021-7026	Dennis Lancet SD510	Duple Dominant	DP31F	1983

7021	EKA221Y	**7023**	EKA223Y	**7024**	EKA224Y	**7025**	EKA225Y	**7026**	EKA226Y

7031	OIW5804	Leyland Tiger TRCTL11/3R	Duple Laser	C49F	1982	Ex Duple demonstrator, 1983
7032	F32ALV	Dennis Javelin 12SDA1907	Duple 320	C50FT	1989	
7033	IIL2503	Dennis Javelin 12SDA1907	Duple 320	C50FT	1989	
7034	J34MKB	Volvo B10M-60	Plaxton Excalibur	C49FT	1992	
7035	J35MKB	Volvo B10M-60	Plaxton Excalibur	C49FT	1992	
7037	HIL5697	Volvo B10M-60	Van Hool Alizée	C49FT	1994	Ex Lee, Langley Moor, 1994
7038	HIL5698	Volvo B10M-60	Van Hool Alizée	C49FT	1994	Ex Lee, Langley Moor, 1994

7201-7250	Volvo B6	Plaxton Pointer	B38F	1994

7201	L201TKA	**7211**	L211TKA	**7221**	L221TKA	**7231**	L231TKA	**7241**	L241TKA
7202	L202TKA	**7212**	L212TKA	**7222**	L222TKA	**7232**	L232TKA	**7242**	L242TKA
7203	L203TKA	**7213**	L213TKA	**7223**	L223TKA	**7233**	L233TKA	**7243**	L243TKA
7204	L204TKA	**7214**	L214TKA	**7224**	L224TKA	**7234**	L234TKA	**7244**	L244TKA
7205	L205TKA	**7215**	L215TKA	**7225**	L225TKA	**7235**	L235TKA	**7245**	L245TKA
7206	L206TKA	**7216**	L216TKA	**7226**	L226TKA	**7236**	L236TKA	**7246**	L246TKA
7207	L207TKA	**7217**	L217TKA	**7227**	L227TKA	**7237**	L237TKA	**7247**	L247TKA
7208	L208TKA	**7218**	L218TKA	**7228**	L228TKA	**7238**	L238TKA	**7248**	L248TKA
7209	L209TKA	**7219**	L219TKA	**7229**	L229TKA	**7239**	L239TKA	**7249**	L249TKA
7210	L210TKA	**7220**	L220TKA	**7230**	L230TKA	**7240**	L240TKA	**7250**	L250TKA

7357	D357JUM	Volkswagen LT55	Optare City Pacer	B25F	1987	Ex Bolton Coachways, 1994
7372	D372JUM	Volkswagen LT55	Optare City Pacer	B25F	1987	Ex Bolton Coachways, 1994
7411	K911OEM	Dennis Dart 9.8SDL3017	Plaxton Pointer	B38F	1993	Ex Blue Triangle, 1994
7455	K955PBG	Dennis Dart 9.8SDL3017	Plaxton Pointer	B36F	1993	Ex Blue Triangle, 1994

7679-7700

Renault-Dodge S56 | Alexander AM | B23F | 1986-87

7679	D679SEM	**7683**	D683SEM	**7687**	D687SEM	**7690**	D690SEM	**7697**	D977TKC
7680	D680SEM	**7684**	D684SEM	**7688**	D688SEM	**7691**	D691SEM	**7698**	D978TKC
7682	D682SEM	**7685**	D685SEM	**7689**	D689SEM	**7693**	D693SEM	**7700**	D700THF

7732	D32KAX	Iveco Daily 49.10	Robin Hood City Nippy	B21F	1986	Ex , 1994
7753	C553BHY	Ford Transit 190	Dormobile	B16F	1986	Ex Woodhead, Morecambe, 1993
7754	C54WBF	Ford Transit 190	Dormobile	B16F	1986	Ex Watson, Lancaster, 1993
7766	D226OOJ	Freight Rover Sherpa	Carlyle	B18F	1986	Ex Bolton Coachways, 1994
7767	D229GLJ	Freight Rover Sherpa	Dormobile	B16F	1986	Ex Bolton Coachways, 1994
7768	D729PTU	Freight Rover Sherpa	Dormobile	B16F	1986	Ex Bolton Coachways, 1994
7777	D137OWG	Renault-Dodge S56	Reeve Burgess	DP25F	1987	Ex Watson, Lancaster, 1993
7778	E178UWF	Renault-Dodge S56	Reeve Burgess Beaver	B25F	1987	Ex Smalley, Heysham, 1993

7801-7810

Optare MetroRider | Optare | B22F* | 1994 | *7802/3 are B17F

7801	K801NTJ	**7803**	K803NTJ	**7805**	K805NTJ	**7807**	L807TFY	**7809**	L809TFY
7802	K802NTJ	**7804**	K804NTJ	**7806**	K806NTJ	**7808**	L808TFY	**7810**	L810TKA

7849	E949GFV	Renault-Dodge S56	Northern Counties	B20F	1987	Ex Bolton Coachways, 1994
7860	D860LND	Renault-Dodge S56	Northern Counties	B20F	1987	Ex GM Buses, 1989
7881	K881UDB	Mercedes-Benz 709D	Plaxton Beaver	B27F	1993	Ex Bolton Coachways, 1994
7883	K883UDB	Mercedes-Benz 709D	Plaxton Beaver	B27F	1993	Ex Bolton Coachways, 1994

7941	G801OVA	Mercedes-Benz 811D	Optare StarRider	B27F	1990	Ex Viscount, 1992
7942	G802OVA	Mercedes-Benz 811D	Optare StarRider	B27F	1990	Ex Viscount, 1992
7943	G803OVA	Mercedes-Benz 811D	Optare StarRider	B27F	1990	Ex Viscount, 1993
7944	G804OVA	Mercedes-Benz 811D	Optare StarRider	B27F	1990	Ex Viscount, 1993

7945-7949

Mercedes-Benz 811D | Marshall | B27F | 1993

7945	K945OEM	**7946**	K946OEM	**7947**	K947OEM	**7948**	K948OEM	**7949**	K949OEM

7950	G950RFL	Mercedes-Benz 811D	Carlyle	B31F	1990	Ex Cambus, 1993
7951	G951RFL	Mercedes-Benz 811D	Carlyle	B31F	1990	Ex Cambus, 1993
7952	G952RFL	Mercedes-Benz 811D	Carlyle	B31F	1990	Ex Cambus, 1993
7953	G953RFL	Mercedes-Benz 811D	Carlyle	B31F	1990	Ex Cambus, 1993
7997	F597FAM	Mercedes-Benz 811D	Optare StarRider	DP31F	1989	Ex Bolton Coachways, 1994
7998	F598FAM	Mercedes-Benz 811D	Optare StarRider	DP31F	1989	Ex Bolton Coachways, 1994

Previous Registrations:

AAX563A	PKG106Y	IIL2503	F33ALV	OIW5804	BCK314Y
ACM768X	DKF683X	JGH118X	KYV335X, 324CLT	UDW639Y	PKG105Y, AAX562A
ACM769X	DKF684X	KHF215P	LJA604P, OIW5804	VKB708	CKC625X
CKB170X	DKF685X	KHG216P	LJA643P, IIL2503	XKC851K	XKC851K, MAN5719
COV8V	FDV824V, KJI3935	L411UFY	L175THF	XKC858K	XKC858K, MAN2692
HIL5697	D637MSJ	L412UFY	L176THF	XKC861K	XKC861K, MAN2133
HIL5698	D638MSJ	LDS190A	JPT906T, 449CLT		

Livery and operations: Cream and crimson (Merseybus, MTL Manchester, Lancashire Travel, Southport and District, Sightseers); yellow, white and black (Merseytravel/Smart); green and yellow (MTL Heysham); green and cream (Merseyrider); silver and maroon (Silver Service); green, blue and white (Bolton Coachways).

Bolton Coachways:	0300, 7357/72, 7693, 7766-8, 7849/81/3, 7997/8.
Fareway:	0001/2, 0030/4/7/40/1, 0093-9, 0108/15/9/21/4/7-9/34-6/8-40/6/9/50/2/4/7, 0159-70/2-8/84, 0213-5, 0377-81/4-7/9-93, 7243-5, 7807-10.
Merseyrider:	1687/9/93-98, 1700-6/8-10/2-5/7/20-5, 1825-33/5-7/9-42, 5202-5/13/87, 6185, 6211, 6301-3, 6939/63, 7246-50, 7411/55, 7945-53..
MTL, Heysham	0526/32/3/53/4/96, 6016/42/62, 6177, 6911/2/27/30/46/69, 7011/2/20, 7753/4/77/8
MTL, Manchester	1105/6/8-13/6, 1726-8/30-2/4-8/40-65, 5208-12/23/7/30/63, 6002/3/5/-10/2/3/20/1/3/5/8/32/3/44/7-9 6051/4/6/61/4/7/70/1/4/5/7/87/8/91-3, 6178-80/6/7/90/2/7, 7220-32.
Sightseers Coaching:	7009/10/3/5/7-9, 7031-5/7/8.
Southport & District:	0261-4, 1035/6, 1449, 1524/51, 1612, 1821-4, 1976/7/83-6, 6107/9/10/24-38, 7201-19, 7944
Wirral Penninsula:	0054/5, 0064-81, 201-19/2/51/65-70, 1067-70, 1472-4/6/7/9/80/2-4/6/8-93, 1507-11/3-8, 1587-93, 1626-8/30-9, 1848/93-1900, 1902-7/21-30/43-50/73-5, 6330-4, 7233-42, 7679/80/2-5/7-91, 7801-6

MERSEYLINE

Meadowhall Ltd, Unit 41 Garston Industrial Estate, Garston, Liverpool, L19 8JB

BNE733N	Daimler Fleetline CRG6LXB	Northern Counties	H43/32F	1974	Ex CMT, Aintree, 1993
GND502N	Daimler Fleetline CRG6LXB	Northern Counties	H43/32F	1974	Ex CMT, Aintree, 1993
NOC585R	Leyland Fleetline FE30AGR	Park Royal	H43/33F	1976	Ex City Fleet, Bootle, 1993
NOC444R	Leyland Fleetline FE30AGR	MCW	H43/33F	1977	Ex City Fleet, Bootle, 1993
NOC445R	Leyland Fleetline FE30AGR	MCW	H43/33F	1977	Ex City Fleet, Bootle, 1993
SDA638S	Leyland Fleetline FE30AGR	Park Royal	H43/33F	1977	Ex J C, Dalton, 1994
SDA503S	Leyland Fleetline FE30AGR	MCW	H43/33F	1977	Ex City Fleet, Bootle, 1993
SDA557S	Leyland Fleetline FE30AGR	MCW	H43/33F	1977	Ex City Fleet, Bootle, 1993
SDA564S	Leyland Fleetline FE30AGR	MCW	H43/33F	1978	Ex Amberley Travel, Leeds, 1993
SDA565S	Leyland Fleetline FE30AGR	MCW	H43/33F	1978	Ex City Fleet, Bootle, 1993
SDA709S	Leyland Fleetline FE30AGR	MCW	H43/33F	1978	Ex City Fleet, Bootle, 1993
SDA767S	Leyland Fleetline FE30AGR	MCW	H43/33F	1978	Ex City Fleet, Bootle, 1993
SDA768S	Leyland Fleetline FE30AGR	MCW	H43/33F	1978	Ex Amberley Travel, Leeds,1993
SDA772S	Leyland Fleetline FE30AGR	MCW	H43/33F	1978	Ex City Fleet, Bootle, 1993
SDA778S	Leyland Fleetline FE30AGR	MCW	H43/33F	1978	Ex J C, Dalton, 1994
SDA784S	Leyland Fleetline FE30AGR	MCW	H43/33F	1978	Ex Amberley Travel, Leeds, 1993

Livery: White and blue

MOUNTAIN GOAT

Mountain Goat Holidays, 10 Victoria Street, Windermere, Cumbria LA23 1AD

F27FEC	Renault Master T35D	Taylor & Hudson	M15	1989
F28FEC	Renault Master T35D	Taylor & Hudson	M15	1989
G218REC	Renault Master T35D	Taylor & Hudson	M15	1990
G963REC	Renault Master T35D	Taylor & Hudson	M15	1990
H487CEC	Renault Master T35D	Taylor & Hudson	M15	1991

Livery: White and green

NIP-ON

K G Hatton, 26 Haywood Gardens, West Park, St Helens, Merseyside, WA10 4JU

PTD672S	Leyland National 11351A/1R		B49F	1978	Ex Bridge, Greenock, 1992
EBM436T	Leyland Leopard PSU5C/4R	Duple Dominant II	C57F	1978	Ex Ashall, Gorton, 1991
EGB93T	Leyland National 11351A/1R		B52F	1979	Ex Western Scottish, 1991
RVN243X	Leyland Leopard PSU3F/4R	Willowbrook 003	DP47F	1982	Ex South Lancs, St Helens, 1992
JNH184Y	Leyland Leopard PSU3G/4R	Eastern Coach Works B51	C46F	1983	Ex United Counties, 1994
A304JFA	Leyland National 2 NL116HLXCT/1R		DP47F	1984	Ex PMT, 1992
A305JFA	Leyland National 2 NL116HLXCT/1R		DP47F	1984	Ex PMT, 1992
A306JFA	Leyland National 2 NL116HLXCT/1R		DP47F	1984	Ex PMT, 1992
D804MNY	Renault-Dodge S56	East Lancashire	B22F	1987	Ex Heaton, Leigh, 1994
D868MDB	Renault-Dodge S56	Northern Counties	B20F	1987	Ex Clydeside 2000, 1994
D893DSF	Renault-Dodge S56	Alexander AM	B25F	1987	Ex Fife Scottish, 1994
F199BCW	Peugeot-Talbot Pullman	Talbot	B20F	1989	Ex Heaton, Leigh, 1994

Livery: Blue and red

Merseyline operate to the Garston area of the city where Leyland Fleetline SDA778S was seen while working service 86 to Liverpool. *Ken Grime*

Nip-On currently run three of six Leyland National 2s Suburban Expresses delivered to PMT in 1992. These feature the high-back seating and red, turquoise and black livery. A306JFA is seen in Corporation Street, St Helens. *Phillip Stephenson*

NORTH WESTERN

North Western Road Car Co, 73 Ormskirk Road, Aintree, Liverpool, Merseyside, L9 5AE
The Bee Line Buzz Company, Liverline Travel Services Limited

A subsidiary of British Bus plc
Depots and outstations: Ormskirk Road, Aintree; Hawthorne Road, Bootle; Hulme Hall Road, Manchester; Miall Street, Rochdale; Mumps, Oldham; Beechwood, Runcorn; Neverstitch Road, Skelmersdale, Athlone Road, Warrington; Wallgate, Wigan and Road Four, Winsford Industrial Estate, Winsford.

3-49 Renault-Dodge S56 | Northern Counties | B22F | 1987 | Ex Ribble, 1989

3	D403NNA	**10**	D410NNA	**17**	D417NNA	**27**	D427NNA	**39**	D439NNA
4	D404NNA	**11**	D411NNA	**18**	D418NNA	**28**	D428NNA	**42**	D442NNA
5	D405NNA	**12**	D412NNA	**19**	D419NNA	**31**	D431NNA	**45**	D445NNA
6	D406NNA	**13**	D413NNA	**21**	D421NNA	**33**	D433NNA	**46**	D446NNA
7	D407NNA	**14**	D414NNA	**24**	D424NNA	**35**	D435NNA	**47**	D447NNA
8	D408NNA	**15**	D415NNA	**25**	D425NNA	**37**	D437NNA	**49**	D449NNA
9	D409NNA	**16**	D416NNA	**26**	D426NNA				

50	J78MHF	Mercedes-Benz 709D	Reeve Burgess Beaver	B29F	1992	Ex Amberline, 1993
51	J734MFY	Mercedes-Benz 709D	Reeve Burgess Beaver	B29F	1992	Ex Amberline, 1993
52	J735MFY	Mercedes-Benz 709D	Reeve Burgess Beaver	B29F	1992	Ex Amberline, 1993

53-67 Mercedes-Benz 709D | Alexander AM | B25F | 1994

53	L153UEM	**56**	L156UEM	**59**	M59WKA	**62**	M62WKA	**65**	M65WKA
54	L154UEM	**57**	M157WWM	**60**	M160WTJ	**63**	M63WKA	**66**	M166WTJ
55	L155UEM	**58**	M158WWM	**61**	M61WKA	**64**	M64WKA	**67**	M67WKA

70-95 Mercedes-Benz 811D | Carlyle | B33F | 1989-90

70	G100TND	**75**	G105TND	**78**	G108TND	**82**	G112TND	**93**	G123TJA
72	G102TND	**76**	G106TND	**79**	G109TND	**88**	G118TND	**94**	G124TJA
73	G103TND	**77**	G107TND	**80**	G110TND	**89**	G119TND	**95**	G125TJA
74	G104TND								

99	H129CDB	Mercedes-Benz 811D	LHE Commuter	B31F	1990	Ex C-Line, 1993
100	H130CDB	Mercedes-Benz 811D	LHE Commuter	B31F	1990	Ex C-Line, 1992

101-105 Mercedes-Benz 709D | Alexander AM | DP23F | 1994

101	M101WKA	**102**	M102WKA	**103**	M103WKA	**104**	M104WKA	**105**	M105WKA

111	D211SKD	Mercedes-Benz L608D	Reeve Burgess	B20F	1986

112-119 Mercedes-Benz L608D | Alexander AM | B20F | 1986

112	D212SKD	**114**	D214SKD	**116**	D216SKD	**118**	D221SKD	**119**	D219SKD
113	D213SKD	**115**	D225SKD	**117**	D217SKD				

139-149 Mercedes-Benz 609D | Reeve Burgess Beaver | B20F* | 1989 | *120-125 are DP19F

139	G139LRM	**142**	G142LRM	**144**	G144LRM	**146**	G146LRM	**148**	G148LRM
140	G140LRM	**143**	G143LRM	**145**	G145LRM	**147**	G147LRM	**149**	G149LRM
141	G141LRM								

150	L150SBG	Dennis Dart 9SDL3034	East Lancashire	B32F	1993
151	L151SBG	Dennis Dart 9SDL3034	East Lancashire	B32F	1993
152	L152SBG	Dennis Dart 9SDL3034	East Lancashire	B32F	1993

Opposite: **Two oerators have come under the North Western umbrella recently, both gaining variations of the North Western livery.** *Above* **is 460, NRN378P in Liverline's version, while** *below* **is 667, GFM110X, in the latest Bee Line colours.** *Phillip Stephenson*

35 HALEWOOD
460
Wallasey (Tunnel)
Manchester (M62)
North & East
Kirkdale
Edge Hill
METROBUS ADVERTISING
Liverline
Liverline
NRN 378P

ARNDALE CENTRE
667
Bee Line
Bee Line

157-170	Dennis Dart 9.8SDL3035	East Lancashire	B40F	1994					
157	M157WKA	**160**	M160WKA	**163**	M163WKA	**166**	M166WKA	**169**	M169WKA
158	M158WKA	**161**	M161WKA	**164**	M164WKA	**167**	M167WKA	**170**	M170WKA
159	M159WKA	**162**	M162WKA	**165**	M165WKA	**168**	M168WKA		

205-213	Leyland National 1151/2R/0401					B52F	1973	Ex Ribble, 1986	
205	PTF733L	**208**	PTF737L	**209**	PTF738L	**212**	PTF755L	**213**	PTF766L
207	PTF735L								

214-255	Leyland National 11351A/1R					B49F	1976-78 Ex Ribble, 1986		
214	SCK687P	**222**	SCK700P	**233**	UHG750R	**241**	ACW764R	**248**	CBV786S
215	SCK688P	**224**	SCK704P	**234**	UHG751R	**242**	ACW765R	**249**	CBV787S
216	SCK689P	**225**	SCK706P	**236**	UHG759R	**243**	CBV766S	**250**	CBV791S
217	SCK690P	**226**	SCK710P	**237**	UHG760R	**244**	CBV767S	**251**	CBV792S
218	SCK692P	**229**	UHG724R	**238**	ACW761R	**245**	CBV769S	**252**	CBV793S
219	SCK693P	**230**	UHG731R	**239**	ACW762R	**246**	CBV773S	**253**	CBV794S
220	SCK698P	**231**	UHG735R	**240**	ACW763R	**247**	CBV774S	**255**	CBV796S
221	SCK699P	**232**	UHG746R						

256	JOX515P	Leyland National 11351A/1R	B49F	1976	Ex Midland Red North, 1988
257	JOX520P	Leyland National 11351A/1R	B49F	1976	Ex Midland Red North, 1988
258	NOE592R	Leyland National 11351A/1R	B49F	1977	Ex Midland Red North, 1988
259	UTU980R	Leyland National 11351A/1R	B49F	1977	Ex Crosville, 1989
260	NOE558R	Leyland National 11351A/1R	B49F	1976	Ex Midland Red North, 1988
262	AFM2W	Leyland National 2 NL116AL11/2R	B52F	1981	Ex Crosville, 1989
263	AFM3W	Leyland National 2 NL116AL11/2R	B52F	1981	Ex Crosville, 1989
264	AFM4W	Leyland National 2 NL116AL11/2R	B52F	1981	Ex Crosville, 1989
267	FCA7X	Leyland National 2 NL116AL11/2R	B52F	1982	Ex Crosville, 1989
275	BVP811V	Leyland National 2 NL116L11/1R	B49F	1980	Ex Midland Red North, 1988
278	EON823V	Leyland National 2 NL116L11/1R	B49F	1980	Ex Midland Red North, 1988
280	FCA10X	Leyland National 2 NL116AL11/2R(6HLXB)	B52F	1982	Ex Crosville, 1989
281	NTU11Y	Leyland National 2 NL116HLXB/2R	B52F	1983	Ex Crosville, 1989
282	NTU12Y	Leyland National 2 NL116HLXB/2R	B52F	1983	Ex Crosville, 1989
283	NTU13Y	Leyland National 2 NL116HLXB/2R	B52F	1983	Ex Crosville, 1989
284	NTU15Y	Leyland National 2 NL116HLXB/2R	B52F	1983	Ex Crosville, 1989
292	NEV681M	Leyland National 11351/1R	B49F	1973	Ex Midland Fox, 1994
293	GHB689N	Leyland National 11351/1R	B52F	1975	Ex Midland Fox, 1994
294	KNV514P	Leyland National 11351/1R	B49F	1976	Ex Midland Fox, 1994
295	PUK638R	Leyland National 11351A/1R	B49F	1977	Ex Midland Fox, 1994
296	PUK640R	Leyland National 11351A/1R	B49F	1977	Ex Midland Fox, 1994
297	HFM177N	Leyland National 11351/1R/SC	B48F	1975	Ex Crosville, 1989
298	HFM186N	Leyland National 11351/1R/SC	B48F	1975	Ex Crosville, 1989
301	DBV845W	Leyland National 2 NL106L11/1R	B44F	1980	Ex Ribble, 1986
303	JCK851W	Leyland National 2 NL106AL11/1R	B44F	1981	Ex Ribble, 1986
306	LFR854X	Leyland National 2 NL106AL11/1R	B44F	1981	Ex Ribble, 1986
309	LFR867X	Leyland National 2 NL106AL11/1R	B44F	1981	Ex Ribble, 1986
310	LFR869X	Leyland National 2 NL106AL11/1R	B44F	1981	Ex Ribble, 1986
314	NWT717M	Leyland National 1151/1R/0401	B52F	1974	Ex Liverline, 1993
315	GUG120N	Leyland National 11351/1R	B52F	1974	Ex Liverline, 1993
316	NUA910P	Leyland National 11351/1R	B52F	1976	Ex Liverline, 1993
317	YFY7M	Leyland National 1151/2R	B49F	1974	Ex Preston, 1990
318	YFY8M	Leyland National 1151/2R	B49F	1974	Ex Preston, 1990
319	YFY1M	Leyland National 1151/2R	B49F	1974	Ex Preston, 1990
320	GHU643N	Leyland National 10351/1R	B44F	1975	Ex Crosville, 1989

322-329	Leyland National 10351/1R*					B41F*	1974-76 Ex London & Country, 1989 *323/7 are B39F; 327 is 10351/1R/SC		
322	LPB207P	**324**	LPB205P	**327**	GPD297N	**328**	LPB209P	**329**	LPB219P
323	LPB195P	**325**	LPB225P						

330-340	Leyland National 10351A/1R					B41F	1976-77 Ex London & Country, 1989		
330	NPK240R	**332**	NFK245R	**335**	UPB301S	**337**	UPB327S	**340**	UPB335S
331	NPK242R	**334**	SPC279R	**336**	UPB322S	**339**	UPB334S		

North Western, Liverline and Bee Line have transferred several vehicles between the three operations, all now sporting liveries applied in a similar style. Photographed approaching Piccadilly, Manchester is 74, G104TND, a Mercedes-Benz 811 with Carlyle bodywork. *Phillip Stephenson*

The latest livery to be applied to 390, G610CFA, is the North Western red and blue scheme. Previously the vehicle wore Happy Days, Midland Red North and Bee Line schemes. This Scania K93 was photographed while working service X5 to Runcorn. *Phillip Stephenson*

341-354 Leyland National 10351B/1R (354 is 6HLXB) B44F 1978-79 Ex Crosville, 1989

341	HMA567T	**344**	JTU583T	**348**	GMB663T	**350**	GMB670T	**352**	MCA678T
342	HMA568T	**345**	EMB643S	**349**	GMB668T	**351**	MCA675T	**354**	ODM680V
343	JTU582T	**347**	GMB657T						

355-373 Leyland National 11351A/1R (6HLX) B49F 1977-79 Ex Crosville, 1990

355	CFM341S	**359**	EMB360S	**363**	GMB379T	**367**	GMB386T	**371**	KMA403T
356	CFM350S	**360**	EMB366S	**364**	GMB380T	**368**	KMA395T	**372**	KMA412T
357	CFM351S	**361**	EMB367S	**365**	GMB384T	**369**	KMA396T	**373**	YTU986S
358	CFM346S	**362**	GMB375T	**366**	GMB385T	**370**	KMA400T		

374	GMB392T	Leyland National 11351A/1R(6HLX)	B49F	1978	Ex C-Line, 1993
375	KMA397T	Leyland National 11351A/1R(6HLX)	B49F	1979	Ex C-Line, 1993

381-388 Dennis Falcon SDA421 East Lancashire EL2000 B48F 1990

381	G381EKA	**383**	G383EKA	**385**	G385EKA	**387**	G387EKA	**388**	G388EKA
382	G382EKA	**384**	G384EKA	**386**	G386EKA				

390	G610CFA	Scania K93CRB	Plaxton Derwent	B57F	1990	Ex Midland Red North, 1993
391	G611CFA	Scania K93CRB	Plaxton Derwent	B57F	1990	Ex Midland Red North, 1993
392	G612CFA	Scania K93CRB	Plaxton Derwent	B57F	1990	Ex Midland Red North, 1993

393	G41HKY	Scania K93CRB	Plaxton Derwent	B57F	1990	Ex Stevensons, 1994
394	D164UTO	Scania K92CRB	East Lancashire	B51F	1986	Ex Stevensons, 1994
395	D163WTV	Scania K92CRB	East Lancashire	B51F	1987	Ex Stevensons, 1994
396	GDM996X	Leyland Leopard PSU5/4R	Willowbrook Warrior (1992)	B52F	1982	Ex Liverline, 1993
397	HEF362N	Leyland Leopard PSU3C/4R	Duple Dominant	B55F	1975	Ex Williamson, Shrewsbury, 1991
398	HFM213J	Bristol RELL6G	Eastern Coach Works	DP22FL	1971	Ex Crosville, 1990
399	790NWC	Leyland National 11351A/1R		DP50FL	1976	Ex Ribble, 1986

403-434

Leyland Atlantean AN68/1R | Park Royal | H43/30F | 1973-74 | Ex Ribble, 1986
415/35 ex Liverline, 1993

403	RTF616M	**407**	RTF621M	**415**	RTF630M	**421**	RTF637M	**432**	RTF648M
404	RTF618M	**408**	RTF622M	**416**	RTF633M	**422**	RTF638M	**434**	RTF651M
405	RTF619M	**411**	RTF625M	**419**	RTF635M	**427**	RTF643M	**435**	RTF652M
406	RTF620M	**414**	RTF629M						

438-459

Leyland Atlantean AN68/1R | Park Royal | H43/30F | 1974 | Ex Ribble, 1986-87

438	GBV123N	**445**	GBV130N	**449**	GBV134N	**454**	GBV139M	**458**	GBV143N
440	GBV125N	**447**	GBV132N	**452**	GBV137N	**457**	NRN400P	**459**	GBV144N
441	GBV126N	**448**	GBV133N	**453**	GBV138M				

460-476

Leyland Atlantean AN68/1R | Park Royal | H43/30F | 1975-76 | Ex Ribble, 1986
... ex Liverline, 1993

460	NRN378P	**464**	NRN388P	**468**	NRN376P	**471**	NRN398P	**474**	NRN389P
461	NRN384P	**465**	NRN373P	**469**	NRN396P	**472**	NRN382P	**475**	NRN403P
462	NRN385P	**466**	NRN392P	**470**	NRN380P	**473**	NRN387P	**476**	NRN399P
463	NRN371P	**467**	NRN394P						

477	JPL134K	Leyland Atlantean PDR1A/1	Park Royal	H43/29D	1972	Ex London & Country, 1990
478	JPL190K	Leyland Atlantean PDR1A/1	Park Royal	H43/29D	1972	Ex London & Country, 1990
479	JPL150K	Leyland Atlantean PDR1A/1	Park Royal	H43/29D	1972	Ex London & Country, 1990
480	JPL136K	Leyland Atlantean PDR1A/1	Park Royal	H43/29D	1972	Ex Bee Line Buzz, 1993

484-491

Leyland Atlantean AN68/1R | Eastern Coach Works | H43/31F | 1976 | Ex Ribble, 1991

484	RFR412P	**487**	RFR409P	**489**	RFR423P	**490**	SFV429P	**491**	SFV433P
486	RFR406P	**488**	RFR422P						

492-497

Leyland Atlantean AN68/2R | Alexander AL | H49/33D* | 1974-75 | Ex Preston, 1987
*494/6/7 are H49/36D

492	GBV102N	**494**	GBV104N	**495**	GBV105N	**496**	GBV106N	**497**	GBV107N
493	GBV103N								

498	RCN104N	Leyland Atlantean AN68/1R	Park Royal	H43/34F	1974	Ex London & Country, 1991
499	RCN99N	Leyland Atlantean AN68/1R	Park Royal	H43/28F	1974	Ex London & Country, 1989
500	TRN466V	Leyland Atlantean AN68A/1R	Eastern Coach Works	H43/31F	1979	Ex Ribble, 1986

501-530

Leyland Atlantean AN68A/1R | Eastern Coach Works | H43/31F | 1980-81 | Ex Ribble, 1986

501	FBV486W	**507**	FBV492W	**513**	FBV490W	**519**	FBV504W	**525**	FBV510W
502	FBV487W	**508**	FBV493W	**514**	FBV499W	**520**	FBV505W	**526**	FBV511W
503	FBV488W	**509**	FBV494W	**515**	FBV500W	**521**	FBV506W	**527**	FBV512W
504	FBV489W	**510**	FBV495W	**516**	FBV501W	**522**	FBV507W	**528**	FBV513W
505	FBV490W	**511**	FBV496W	**517**	FBV502W	**523**	FBV508W	**529**	FBV514W
506	FBV491W	**512**	FBV497W	**518**	FBV503W	**524**	FBV509W	**530**	FBV515W

531	RFR414P	Leyland Atlantean AN68/1R	Eastern Coach Works	H43/31F	1976	Ex Ribble, 1993
532	SFV427P	Leyland Atlantean AN68/1R	Eastern Coach Works	H43/31F	1976	Ex Bee Line Buzz, 1993
533	SFV428P	Leyland Atlantean AN68/1R	Eastern Coach Works	H43/31F	1976	Ex Bee Line Buzz, 1993
534	SFV432P	Leyland Atlantean AN68/1R	Eastern Coach Works	H43/31F	1976	Ex Ribble, 1993
535	SFV436P	Leyland Atlantean AN68/1R	Eastern Coach Works	H43/31F	1976	Ex Ribble, 1993
536	NNO65P	Leyland Atlantean AN68A/1R	Eastern Coach Works	H43/31F	1976	Ex Bee Line Buzz, 1993
537	TPU70R	Leyland Atlantean AN68A/1R	Eastern Coach Works	H43/31F	1977	Ex Colchester, 1992
538	TPU72R	Leyland Atlantean AN68A/1R	Eastern Coach Works	H43/31F	1977	Ex C-Line, 1993

The variety of double-deck vehicles in the North Western fleet was contributed to by the many inherited from fleets taken over, or purchased for the expansion into Altrincham. *Top* is Leyland Atlantean 421, RTF637M, new to Ribble was in Liverline colours when pictured in Liverpool, bound for Netherton. *Centre:* Alexander-bodied Leyland Atlantean 494, GBV104N, is seen in Liverpool heading for Thornton. These were formerly with Preston and are unusual for the North Western fleet in being dual-doored. *Below:* Typical of the former NBC fleets is the Bristol VR with Eastern Coach Works bodies. While the Ribble examples stayed in Lancashire during the formation of North Western, several examples arrived in the fleet with the Crosville depots of Northwich and Runcorn. Photographed leaving Crewe is 549, WTU499W.
Mike Fowler/ Phillip Stephenson/ Cliff Beeton

543-563 Bristol VRT/SL3/501 | Eastern Coach Works | H43/31F* | 1978-81 | Ex Crosville, 1989
*545/7/49/50/53/63 are H43/30F

543	RMA437V	547	PCA425V	551	FTU382T	555	ODM410V	559	RMA441V
544	PCA422V	548	RLG427V	552	FTU386T	556	RMA436V	560	WTU492W
545	PCA423V	549	WTU499W	553	FTU391T	557	RMA438V	562	WTU496W
546	PCA424V	550	BTU374S	554	JMB406T	558	RMA439V	563	WTU497W

564-568 Bristol VRT/SL3/6LXB | Eastern Coach Works | H43/31F* | 1980 | Ex Crosville, 1989
*566/8 are H43/30F

564	VCA453W	565	VCA454W	566	VCA455W	567	VCA456W	568	VCA463W

569	WTU479W	Bristol VRT/SL3/6LXC(6LXB)	Eastern Coach Works	H43/30F	1981	Ex Crosville, 1989
570	WTU480W	Bristol VRT/SL3/6LXC(6LXB)	Eastern Coach Works	H43/30F	1981	Ex Crosville, 1989
573	GND495N	Leyland Atlantean AN68/1R	Northern Counties	H43/32F	1974	Ex London & Country, 1989
574	GND496N	Leyland Atlantean AN68/1R	Northern Counties	H43/32F	1974	Ex London & Country, 1989
575	JVM989N	Leyland Atlantean AN68A/1R	Northern Counties	H43/32F	1975	Ex Ribble, 1992
576	HNB31N	Leyland Atlantean AN68/1R	Northern Counties	H43/32F	1975	Ex London & Country, 1989
577	HNB33N	Leyland Atlantean AN68/1R	Northern Counties	H43/32F	1975	Ex London & Country, 1989
578	HNB37N	Leyland Atlantean AN68/1R	Northern Counties	H43/32F	1975	Ex London & Country, 1989
579	JDB117N	Leyland Atlantean AN68A/1R	Northern Counties	H43/32F	1975	Ex Ribble, 1992
580	GND499N	Leyland Atlantean AN68/1R	Northern Counties	H43/32F	1974	Ex London & Country, 1989

581-594 Leyland Atlantean AN68A/1R | Northern Counties | H43/32F | 1975-76 | Ex Ribble, 1991-93

581	JND984N	584	JDB114N	587	KBU913P	590	LJA601P	593	LJA629P
582	JND985N	585	JDB115N	588	JND917P	591	LJA616P	594	KDB684P
583	JDB112N	586	JDB122N	589	KDB685P	592	LJA620P		

600-625 Leyland Olympian ONLXB/1R | Eastern Coach Works | H45/32F | 1983-85 | Ex Ribble, 1986

600	DBV133Y	605	A141MRN	610	B149TRN	615	B155TRN	621	B965WRN
601	DBV135Y	606	A144OFR	611	B150TRN	616	B960WRN	622	B966WRN
602	DBV136Y	607	A146OFR	612	B151TRN	618	B962WRN	623	B967WRN
603	A139MRN	608	A147OFR	613	B153TRN	619	B963WRN	624	B968WRN
604	A140MRN	609	B148TRN	614	B154TRN	620	B964WRN	625	B969WRN

626	G626EKA	Dennis Dominator DDA1031	East Lanacshire	H47/29F	1990
627	G627EKA	Dennis Dominator DDA1031	East Lanacshire	H47/29F	1990
628	G628EKA	Dennis Dominator DDA1031	East Lanacshire	H47/29F	1990
629	G629EKA	Dennis Dominator DDA1031	East Lanacshire	H47/29F	1990

630-635 Dennis Dominator DDA1026 | East Lanacshire | DPH43/25F | 1989

630	F630BKD	632	F632BKD	633	F633BKD	634	F634BKD	635	F635BKD
631	F631BKD								

636	F636BKD	Dennis Dominator DDA1025	East Lanacshire	H45/31F	1989
637	F637BKD	Dennis Dominator DDA1025	East Lanacshire	H45/31F	1989
638	F638BKD	Dennis Dominator DDA1025	East Lanacshire	H45/31F	1989
639	F639BKD	Dennis Dominator DDA1025	East Lanacshire	H45/31F	1989

640-653 Volvo Citybus B10M-50 | East Lancashire | H49/39F | 1989-90

640	G640CHF	643	G643CHF	646	G661DTJ	649	G649EKA	652	G652EKA
641	G641CHF	644	G659DTJ	647	G647EKA	650	G650EKA	653	G653EKA
642	G642CHF	645	G660DTJ	648	G648EKA	651	G651EKA		

654-662 Leyland Olympian ONLXB/1R | Eastern Coach Works | H45/32F | 1983-84 | Ex Crosville, 1989

654	PFM126Y	656	PFM129Y	658	A140SMA	660	A149UDM	662	A153UDM
655	PFM128Y	657	A139SMA	659	A141SMA	661	A151UDM		

663	G663FKA	Dennis Dominator DDA1031	East Lanacshire	H47/29F	1990	
664	G664FKA	Dennis Dominator DDA1031	East Lanacshire	H47/29F	1990	
665	G665FKA	Dennis Dominator DDA1031	East Lanacshire	H47/29F	1990	
666	G667FKA	Dennis Dominator DDA1031	East Lanacshire	H47/29F	1990	
667	GFM110X	Leyland Olympian ONLXB/1R	Eastern Coach Works	H45/32F	1982	Ex Crosville, 1989
668	A142SMA	Leyland Olympian ONLXB/1R	Eastern Coach Works	H45/32F	1983	Ex Crosville, 1989
669	A148SMA	Leyland Olympian ONLXB/1R	Eastern Coach Works	H45/32F	1983	Ex Crosville, 1989
670	A147SMA	Leyland Olympian ONLXB/1R	Eastern Coach Works	H45/32F	1983	Ex Crosville, 1989

Liverline operated ten Scania N113 double deck buses, all with Northern Counties bodywork. Photographed in Liverpool about to head south to Halewood is the first of the batch, 671, G34HKY, with 673, G36HKY parked behind. *Cliff Beeton*

An interesting arrival from Graham's of Paisley is 700, YDS651S, one of the Bee Line liveried buses allocated to Rochdale. It is seen leaving that town on the main service to Ashton-under-Lyne, the 409. *Roy Marshall*

671-680		Scania N113DRB		Northern Counties		H47/33F	1990-91	Ex Liverline, 1993	
671	G34HKY	**673**	G36HKY	**675**	G38HKY	**677**	G714LKW	**679**	H804RWJ
672	G35HKY	**674**	G37HKY	**676**	G711LKW	**678**	H803RWJ	**680**	H805RWJ

700	YDS651S	Leyland Atlantean AN68A/1R	Alexander AD	H45/33F	1977	Ex Graham's, Paisley, 1990
701	XJA531L	Leyland Atlantean AN68/1R	Park Royal	H43/32F	1973	Ex Ribble, 1989
702	PUF136M	Leyland Atlantean AN68/1R	Park Royal	H43/30F	1974	Ex Liverline, 1993
703	PUF718M	Leyland Atlantean AN68/1R	Park Royal	H43/30F	1974	Ex Ribble, 1989
704	PUF721M	Leyland Atlantean AN68/1R	Park Royal	H43/30F	1974	Ex Ribble, 1989
705	GDR211N	Leyland Atlantean AN68/1R	Park Royal	H43/30F	1975	Ex Plymouth, 1989
707	GDR213N	Leyland Atlantean AN68/1R	Park Royal	H43/30F	1975	Ex Plymouth, 1989
708	GDR214N	Leyland Atlantean AN68/1R	Park Royal	H43/30F	1975	Ex Plymouth, 1989
709	SCD728N	Leyland Atlantean AN68/1R	Park Royal	H43/30F	1974	Ex London & Country, 1989
710	SCD730N	Leyland Atlantean AN68/1R	Park Royal	H43/30F	1974	Ex London & Country, 1989
711	SCD731N	Leyland Atlantean AN68/1R	Park Royal	H43/30F	1974	Ex London & Country, 1989
712	SCD733N	Leyland Atlantean AN68/1R	Park Royal	H43/30F	1974	Ex London & Country, 1989
713	LCD44P	Leyland Atlantean AN68/1R	Roe	H43/30F	1976	Ex Ribble, 1989
714	SCD729N	Leyland Atlantean AN68/1R	Park Royal	H43/30F	1974	Ex London & Country, 1989
715	LCD42P	Leyland Atlantean AN68/1R	Roe	H43/30F	1976	Ex Ribble, 1989
716	LCD46P	Leyland Atlantean AN68/1R	Roe	H43/30F	1976	Ex Ribble, 1989
718	LTK92R	Leyland Atlantean AN68A/1R	Roe	H43/30F	1977	Ex Hyndburn, 1990
719	LTK98R	Leyland Atlantean AN68A/1R	Roe	H43/30F	1977	Ex Plymouth, 1990

720	OCO107S	Leyland Atlantean AN68A/1R	Roe	H43/30F	1978	Ex Plymouth, 1990
721	UPK125S	Leyland Atlantean AN68A/1R	Park Royal	H43/30F	1978	Ex London & Country, 1991
722	UPK130S	Leyland Atlantean AN68A/1R	Park Royal	H43/30F	1978	Ex London & Country, 1991
723	UPK139S	Leyland Atlantean AN68A/1R	Park Royal	H43/30F	1978	Ex London & Country, 1991
724	UPK141S	Leyland Atlantean AN68A/1R	Park Royal	H43/30F	1978	Ex London & Country, 1991
725	OCO109S	Leyland Atlantean AN68A/1R	Roe	H43/30F	1978	Ex Plymouth, 1990
726	UPK124S	Leyland Atlantean AN68A/1R	Park Royal	H43/30F	1978	Ex London & Country, 1991
727	UPK137S	Leyland Atlantean AN68A/1R	Park Royal	H43/30F	1978	Ex London & Country, 1991
728	UPK138S	Leyland Atlantean AN68A/1R	Park Royal	H43/30F	1978	Ex London & Country, 1991
729	UPK140S	Leyland Atlantean AN68A/1R	Park Royal	H43/30F	1978	Ex London & Country, 1991
730	VPA150S	Leyland Atlantean AN68A/1R	Park Royal	H43/30F	1978	Ex London & Country, 1991
731	VPA154S	Leyland Atlantean AN68A/1R	Park Royal	H43/30F	1978	Ex Luton & District, 1992
732	VPA156S	Leyland Atlantean AN68A/1R	Park Royal	H43/30F	1978	Ex Luton & District, 1992
733	VPA157S	Leyland Atlantean AN68A/1R	Park Royal	H43/30F	1978	Ex Luton & District, 1992
734	VPA159S	Leyland Atlantean AN68A/1R	Park Royal	H43/30F	1978	Ex Luton & District, 1992
735	XPG162T	Leyland Atlantean AN68A/1R	Park Royal	H43/30F	1978	Ex Luton & District, 1992
736	XPG163T	Leyland Atlantean AN68A/1R	Park Royal	H43/30F	1978	Ex Luton & District, 1992
737	XPG166T	Leyland Atlantean AN68A/1R	Park Royal	H43/30F	1978	Ex Luton & District, 1992
738	XPG167T	Leyland Atlantean AN68A/1R	Park Royal	H43/30F	1978	Ex Luton & District, 1992
739	XPG169T	Leyland Atlantean AN68A/1R	Park Royal	H43/30F	1978	Ex Luton & District, 1992
740	XPG170T	Leyland Atlantean AN68A/1R	Park Royal	H43/30F	1978	Ex Luton & District, 1992
741	XPG171T	Leyland Atlantean AN68A/1R	Park Royal	H43/30F	1978	Ex Luton & District, 1992
742	XPG177T	Leyland Atlantean AN68A/1R	Roe	H43/30F	1978	Ex London & Country, 1991
743	XPG178T	Leyland Atlantean AN68A/1R	Roe	H43/30F	1978	Ex London & Country, 1991
744	XPG181T	Leyland Atlantean AN68A/1R	Roe	H43/30F	1978	Ex London & Country, 1991
745	XPG193T	Leyland Atlantean AN68A/1R	Roe	H43/30F	1979	Ex London & Country, 1991
746	XPG176T	Leyland Atlantean AN68A/1R	Roe	H43/30F	1978	Ex London & Country, 1991
747	XPG189T	Leyland Atlantean AN68A/1R	Roe	H43/30F	1979	Ex London & Country, 1991
748	XPG198T	Leyland Atlantean AN68A/1R	Roe	H43/30F	1979	Ex Luton & District, 1992
749	XPG202T	Leyland Atlantean AN68A/1R	Roe	H43/30F	1979	Ex Luton & District, 1992
750	EPH230V	Leyland Atlantean AN68A/1R	Roe	H43/30F	1980	Ex London & Country, 1991
761	NOC381R	Leyland Fleetline FE30ALR(6LXB)	MCW	H43/33F	1976	Ex Amberline, 1993
762	NOC382R	Leyland Fleetline FE30ALR(6LXB)	MCW	H43/33F	1976	Ex Amberline, 1993
763	NOC390R	Leyland Fleetline FE30ALR(6LXB)	MCW	H43/33F	1976	Ex Amberline, 1993
765	NOC421R	Leyland Fleetline FE30AGR	MCW	H43/33F	1976	Ex Amberline, 1993
766	NOC436R	Leyland Fleetline FE30AGR	MCW	H43/33F	1977	Ex Amberline, 1993
776	776DYE	AEC Routemaster 5RM	Park Royal	H36/28R	1963	Ex GM Buses, 1991
818	OMA510V	Leyland Leopard PSU3E/4R	Duple Dominant II Express	C49F	1979	Ex Crosville, 1990
819	A39SMA	Leyland Tiger TRCTL11/2RH	Duple Laser	C49F	1984	Ex Crosville, 1989
820	A40SMA	Leyland Tiger TRCTL11/2RH	Duple Laser	C49F	1984	Ex Crosville, 1989
821	A41SMA	Leyland Tiger TRCTL11/2RH	Duple Laser	C49F	1984	Ex Crosville, 1989
823	1205FM	Leyland Tiger TRCTL11/2RH	Duple Laser	C49F	1984	Ex Crosville, 1989
825	B148ALG	Leyland Tiger TRCTL11/2RH	Duple Laser 2	C49F	1984	Ex Crosville, 1989
826	B149ALG	Leyland Tiger TRCTL11/2RH	Duple Laser 2	C49F	1984	Ex Crosville, 1989
827	B150ALG	Leyland Tiger TRCTL11/2RH	Duple Laser 2	C49F	1984	Ex Crosville, 1989
835	LOA835X	Leyland Leopard PSU3F/3R	Willowbrook 003	DP49F	1982	Ex Midland Fox, 1993
836	LOA836X	Leyland Leopard PSU3F/3R	Willowbrook 003	DP49F	1982	Ex Midland Fox, 1993
837	LOA837X	Leyland Leopard PSU3F/3R	Willowbrook 003	DP49F	1982	Ex Midland Fox, 1993
853	L153UKB	Dennis Dart 9SDL3034	Plaxton Pointer	B20F	1994	
854	L154UKB	Dennis Dart 9SDL3034	Plaxton Pointer	B20F	1994	
855	L155UKB	Dennis Dart 9SDL3034	Plaxton Pointer	B20F	1994	
856	L156UKB	Dennis Dart 9SDL3034	Plaxton Pointer	B20F	1994	
857	G644EVN	CVE Omni	CVE	B15FL	1990	Ex Greater Manchester PTE, 1994
860	G238GCC	Iveco Daily 49.10	Carlyle Dailybus 2	B25F	1989	On loan
863	G245GCC	Iveco Daily 49.10	Carlyle Dailybus 2	B25F	1989	
866	G250GCC	Iveco Daily 49.10	Carlyle Dailybus 2	B25F	1989	On loan
871	C460SJU	Ford Transit 190	Dormobile	B16F	1986	Ex Midland Fox, 1994
872	C510TJF	Ford Transit 190	Dormobile	B16F	1986	Ex Midland Fox, 1994
873	C568TUT	Ford Transit 190	Dormobile	B16F	1986	Ex Midland Fox, 1994
881	838AFM	Bristol Lodekka LD6G	Eastern Coach Works	H33/27R	1957	Ex Driver trainer, 1993
916	VDB916	Leyland Leopard PSU3/3RT	Alexander AY	DP49F	1962	Acquired 1994
935	RWG453N	Leyland Leopard PSU3/3R	Alexander AY	B53F	1974	Acquired 1994

Previous Registrations:

1205FM	A43SMA	790NWC	SCK691P	GDM996X	VUR217J

Livery: Red, blue, white and grey; red, yellow, white and grey (Bee Line); pale and dark blue, white and grey (Liverline)

In 1993 five Leyland Fleetlines, all with MCW bodywork, came from the Amberline fleet. One of these, 763, NOC390R, is seen in Liverpool heading out of the city towards Netherly. *Ken Grime*

Purchased for a particular service in Wigan were four Dennis Darts with Plaxton Pointer bodywork. These are painted in a distinctive Easylink scheme using silver, red and orange. Photographed at the bus station is 856, L156UKB. *Roy Marshall*

OGDEN'S

Ogden Coaches, 204 Clipsley Lane, Haydock, St Helens, Merseyside, WA11 0HD

YCW847N	Bristol RESL6G	East Lancashire	B46F	1974	Ex Nip-On, St Helens, 1993
YCW849N	Bristol RESL6G	East Lancashire	B46F	1974	Ex Burnley & Pendle, 1991
JIW374	Leyland Leopard PSU3E/4R	Duple Dominant II	C46F	1978	Ex West Riding, 1988
HFF234	AEC Reliance 6U3ZR	Plaxton Supreme IV	C53F	1979	Ex Harding, Halebank, 1984
JUM977V	Ford R1114	Duple Dominant II	C53F	1979	Ex Wray, Summerbridge, 1983
GHB221W	Leyland National 2 NL106L11/1R		B41F	1980	Ex National Welsh, 1990
GHB223W	Leyland National 2 NL106L11/1R		B41F	1980	Ex National Welsh, 1990
GHB225W	Leyland National 2 NL106L11/1R		B41F	1980	Ex National Welsh, 1990
GHB226W	Leyland National 2 NL106L11/1R		B41F	1980	Ex National Welsh, 1990
JWO232W	Leyland National 2 NL116L11/1R		DP44F	1980	Ex National Welsh, 1990
OIJ2645	Leyland Leopard PSU3E/4R	Duple Dominant IV	C49F	1982	Ex Mercer, Longridge, 1987
HIL6956	Leyland Tiger TRCTL11/2R	Plaxton Supreme V	C46FT	1982	Ex SUT, Sheffield, 1987
DOI9172	DAF SB2300DHS585	Plaxton Paramount 3200 II	C53F	1986	
HIL2381	DAF MB230DKVL615	Van Hool Alizée	C55F	1987	Ex Ardenvale, Knowle, 1991
A302JFA	Leyland National 2 NL116HLXCT/1R		DP47F	1984	Ex PMT, 1993
C208TLY	Renault-Dodge S56	Reeve Burgess	M8	1985	Ex Hertz, Streatham, 1993
E909EAY	DAF MB230DKFL615	Plaxton Paramount 3200 III	C57F	1987	Ex Dennis's, Ashton-u-Lyne, 1994
E759HJF	Mercedes-Benz 307D	Yeates	M12	1988	
470DOT	DAF MB230LB615	Plaxton Paramount 3500 III	C53F	1988	Ex Smith, Alcester, 1990
F215RJX	DAF SB2300DHS585	Plaxton Paramount 3500 III	C53F	1988	
F660OHD	DAF SB2305DHTD585	Plaxton Paramount 3200 III	C53F	1988	Ex Sweyne, Swinefleet, 1992
F848YJX	DAF SB220LC550	Optare Delta	B49F	1989	
F849YJX	DAF SB220LC550	Optare Delta	DP48F	1989	
G228HCP	DAF MB230LT615	Plaxton Paramount 3500 III	C53F	1990	
H533YCX	DAF SB220LC550	Ikarus Citi Bus	B48F	1991	Ex OK, Bishop Auckland, 1992
H539YCX	DAF SB220LC550	Ikarus Citi Bus	B48F	1991	Ex OK, Bishop Auckland, 1992
J791EAY	DAF SB2305DKV601	Van Hool Alizée	C51F	1992	Ex Central, Birmingham, 1994
K30GGY	Iveco Daily 49.10	Marshall C29	B23F	1993	
K333DOT	DAF SB3000DKVF601	Van Hool Alizée	C51FT	1993	
L10GGY	Ford Transit VE6	Deansgate	M14	1993	

Previous Registrations:

470DOT	E641KCX	HIL2381	D618YCX	K333DOT	K540RJX
DOI9172	C631TUT, COI6771	HIL6956	NHL261X, JOI2949	OIJ2645	RHG911X
HFF234	DCH359T	JIW374	VKU106S		

Livery: Red and grey; white, red and grey (coaches).

One of a pair of Optare Deltas in the Ogdens fleet is F849YJX. Two vehicles with similar chassis but Ikarus bodywork are also operated.
Phillip Stephenson

PIONEER

J M Whitehead, 10 Fairfax Drive, Smithybridge, Littleborough, Lancashire, OL15 8RA

Depot: Scot Industrial Park, Fishwick Street, Rochdale.

	CWG286	Leyland Tiger PS1/1	Alexander	C35F	1950	Ex Bruce, Overtown, 1993
	RDV419H	Bristol RELH6G	Eastern Coach Works	DP45F	1970	Ex Southern National, 1994
w	289CLT	Leyland Leopard PSU3E/4R	Duple Dominant	C53F	1973	Ex Bradshaw, Middleton, 1992
w	YSD814T	Seddon Pennine 7	Alexander AY	DP49F	1978	Ex Western Scottish, 1991
	SSX601V	Seddon Pennine 7	Alexander AYS	B53F	1980	Ex Mayne, Manchester, 1993
	HSC161X	Leyland Cub CU435	Duple Dominant	B31F	1981	Ex Lothian, 1991
	HSC174X	Leyland Cub CU435	Duple Dominant	B31F	1981	Ex Lothian, 1991
	MSL155X	Leyland Leopard PSU3F/4RT	Willowbrook 003	C49F	1981	Ex Clydeside 2000, 1994
	RPB955X	Leyland Leopard PSU3G/4R	Eastern Coach Works B51	C49F	1982	Ex Stagecoach South, 1994
	C41HDT	Dennis Domino SDA1201	Optare	B33F	1985	Ex Stevensons, 1994
	C46HDT	Dennis Domino SDA1201	Optare	B33F	1985	Ex Stevensons, 1994
	D972TKC	Renault-Dodge S56	Northern Counties	B22F	1987	Ex Merseybus, 1992
	D973TKC	Renault-Dodge S56	Northern Counties	B22F	1987	Ex Merseybus, 1992
	KUX774	Volkswagen LT55	Optare City Pacer	B25F	1987	Ex Kingston-upon-Hull, 1993
	G48VVM	Mercedes-Benz 609D	Made-to-Measure	DP24F	1990	
	K367TJF	Mercedes-Benz 811D	Dormobile Routemaker	B33F	1992	

Previous Registrations:

289CLT	FAU48L	KUX774	E941WWE
CWG286	From new	MSL155X	PRM629X, 588SC

Livery: Purple and yellow.

Photographed in Rochdale, Pioneer C41HDT is one of a pair of Optare-bodied Dennis Dominoes new to West Yorkshire PTE in this fleet. This particular vehicle came from Stevensons. *Ken Grime*

POWERCRAFTS

J A Knighton, Unit 1 Ciceley Bridge Mill, Ciceley Street, Blackburn, BB1 1EW

SXI9038	Leyland Tiger TRCTL11/3R	Duple Laser	C46FT	1982	Ex Stuart, Hyde, 1992
C564TUT	Ford Transit 190	Dormobile	B16F	1988	Ex Brown, Oswaldtwistle, 1994
D856KWR	Freight Rover Sherpa	Dormobile	B20F	1987	Ex Yorkshire Rider, 1991
D61NOF	Freight Rover Sherpa	Carlyle	B18F	1987	Ex Jump, Blackburn, 1993
D207NON	Freight Rover Sherpa	Carlyle	B18F	1987	Ex Bee Line Buzz, 1991
E144RAX	Freight Rover Sherpa	Carlyle Citybus 2	B20F	1987	Ex Red & White, 1993
E156RNY	Freight Rover Sherpa	Carlyle Citybus 2	B20F	1987	Ex Red & White, 1993
E817SUM	Volkswagen LT55	Optare City Pacer	DP25F	1988	Ex M&E, Clayton-le-Dale, 1990
F836BCW	Mercedes-Benz 811D	Reeve Burgess Beaver	B33F	1989	
F515SCW	Mercedes-Benz 407D	Reeve Burgess	M15	1989	Ex Duckworth, Gisburn, 1990

Previous Registrations:
SXI9038 SSJ134Y, WLT917, PSP491Y

Livery: Red, white and black

Seen passing Openshaw Drive in Blackburn is Powercraft D856KWR. The service 12 is a local one from the town's bus station. *Ken Grime*

PRESTON

Prestonbus Ltd, 221 Deepdale Road, Preston,
Lancashire, PR1 6NY

1	DRN1Y	Leyland Atlantean AN68D/2R	East Lancashire	DPH45/29F	1983	
2	DRN2Y	Leyland Atlantean AN68D/2R	East Lancashire	DPH45/29F	1983	
6	D456BEO	Renault-Dodge S56	East Lancashire	B22F	1986	Ex Ribble, 1991
7	D32SAO	Renault-Dodge S56	Reeve Burgess	B23F	1986	Ex Cumberland, 1991
8	D458BEO	Renault-Dodge S56	East Lancashire	DP22F	1986	Ex Ribble, 1991
9	D459BEO	Renault-Dodge S56	East Lancashire	DP22F	1986	Ex Ribble, 1991
10	F210YHG	Leyland Lynx LX2L10ZR1R	Leyland Lynx	B47F	1989	
11	F211YHG	Leyland Lynx LX2L10ZR1R	Leyland Lynx	B47F	1989	
12	F212YHG	Leyland Lynx LX2L10ZR1R	Leyland Lynx	DP45F	1989	
13	F213YHG	Leyland Lynx LX2L10ZR1R	Leyland Lynx	DP45F	1989	

14-18 Leyland Lynx LX2R11C15Z4R Leyland Lynx DP45F 1989

14	G214KRN	15	G215KRN	16	G216KRN	17	G217KRN	18	G218KRN

23-29 Leyland Lynx LX2R11C15Z4R Leyland DP45F 1990

23	H23YBV	26	H26YBV	27	H27YBV	28	H28YBV	29	H29YBV
24	H24YBV								

32	F32AHG	Leyland Olympian ONCL10/2RZ	Northern Counties	H51/34F	1989
33	A33MRN	Leyland Olympian ONTL11/2R	Eastern Coach Works	H47/25F	1984
34	G34OCK	Leyland Olympian ONCL10/1RZ	Eastern Coach Works	DPH43/29F	1990
35	G35OCK	Leyland Olympian ONCL10/1RZ	Eastern Coach Works	DPH43/29F	1990
36	G36OCK	Leyland Olympian ONCL10/1RZ	Eastern Coach Works	H47/31F	1990
37	G37OCK	Leyland Olympian ONCL10/1RZ	Eastern Coach Works	H47/31F	1990
40	D40AFV	Leyland Tiger TRBTL11/2RP	Duple 320	C51F	1987

Sole coach in the Preston fleet is Leyland Tiger 40, D40AFV. Fitted with Duple 320 coachwork, it can frequently be found on excursions and private hire duties as was the case when photographed earlier this year. *Phillip Stephenson*

1989 saw the arrival of nine Leyland Lynx, some Renault-Dodge S56 minibuses and a single Leyland Olympian. Numbered 32, F32AHG is fitted with Northern Counties bodywork, seen here leaving the bus station in Preston, complete with a Sainsbury's advert, bound for the Asda Superstore in the Sherwood district of the town. *Roy Marshall*

41-74 Renault-Dodge S56* Northern Counties B22F* 1987/88 *68/9 are type S46
*44-6/67 are B20F; 47-49 are B25F; 70/71 are DP20F

41	D41AFV	**48**	E48MCK	**55**	D755YCW	**62**	D762YCW	**69**	D769YCW
42	D42AFV	**49**	E49MCK	**56**	D756YCW	**63**	D763YCW	**70**	D870ABV
43	D43AFV	**50**	D750YCW	**57**	D757YCW	**64**	D764YCW	**71**	D871ABV
44	E44FFV	**51**	D751YCW	**58**	D758YCW	**65**	D765YCW	**72**	D72AFV
45	E45GRN	**52**	D752YCW	**59**	D759YCW	**66**	D766YCW	**73**	D73AFV
46	E46GRN	**53**	D753YCW	**60**	D760YCW	**67**	D767YCW	**74**	D74AFV
47	E47KBV	**54**	D754YCW	**61**	D761YCW	**68**	D768YCW		

75-92 Renault-Dodge S56 Northern Counties B25F 1988/89

75	E75LFR	**79**	E79MHG	**83**	E83MHG	**87**	E87MHG	**90**	F90UHG
76	E76LFR	**80**	E80MHG	**84**	E84MHG	**88**	F88UHG	**91**	F91AHG
77	E77LFR	**81**	E81MHG	**85**	E85MHG	**89**	F89UHG	**92**	F92AHG
78	E78MHG	**82**	E82MHG	**86**	E86MHG				

101	H101BFR	Leyland Olympian ON2R50C13Z4	Northern Counties	H47/30F	1991	
102	H102BFR	Leyland Olympian ON2R50C13Z4	Northern Counties	H47/30F	1991	
103	H103BFR	Leyland Olympian ON2R50C13Z4	Northern Counties	H47/30F	1991	
104	H104BFR	Leyland Olympian ON2R50C13Z4	Northern Counties	H47/30F	1991	
106	J976PRW	Leyland Olympian ON2R50C13Z4	Leyland	H47/30F	1991	Ex Volvo bus demonstrator, 1992

107-114 Leyland Olympian ON2R50C13Z4 Leyland H47/31F* 1991-92 *107/14 are DPH43/29F

107	J107KCW	**109**	J109KCW	**112**	J112KCW	**113**	J113KCW	**114**	J114KCW
108	J108KCW	**110**	J110KCW						

In 1991, Preston ordered eight all-Leyland Olympians, one of which was the then current demonstrator. The seven new vehicles were assembled at Lillyhall, with two being fitted with high-back seating. Fitted with traditional seating is 108, J108KCW, photographed in Preston.
Roy Marshall

141-150		Leyland Atlantean AN68A/2R		Alexander AL		H49/36F	1980		
141	UHG141V	**143**	UHG143V	**145**	UHG145V	**147**	UHG147V	**149**	UHG149V
142	UHG142V	**144**	UHG144V	**146**	UHG146V	**148**	UHG148V	**150**	UHG150V

151-157		Leyland Atlantean AN68B/2R		East Lancashire		H50/36F	1981		
151	GFV151W	**153**	GFV153W	**155**	GFV155W	**156**	GFV156W	**157**	GFV157W
152	GFV152W	**154**	GFV154W						

158-165		Leyland Atlantean AN68C/2R		East Lancashire		H50/36F	1982		
158	OBV158X	**160**	OBV160X	**162**	OBV162X	**164**	OBV164X	**165**	OBV165X
159	OBV159X	**161**	OBV161X	**163**	OBV163X				

166-177		Leyland Atlantean AN68D/2R		East Lancashire		H50/36F	1982-83		
166	URN166Y	**169**	URN169Y	**172**	URN172Y	**174**	DRN174Y	**176**	DRN176Y
167	URN167Y	**170**	URN170Y	**173**	DRN173Y	**175**	DRN175Y	**177**	DRN177Y
168	URN168Y	**171**	URN171Y						

Livery: Blue and ivory

RELIANCE TRAVEL

D W Henshall & H Varley, 157 Oldham Road, Ashton-u-Lyne,
Greater Manchester, OL7 9AR

YNA327M	Daimler Fleetline CRG6LXB	Northern Counties	H43/32F	1974	Ex KMP, Llanberis, 1993
YNA358M	Daimler Fleetline CRG6LXB	Northern Counties	H43/32F	1974	Ex East Yorkshire, 1993
GHM782N	Daimler Fleetline CRL6	MCW	H44/32F	1974	Ex Bennett, Warrington, 1994
TDK291S	AEC Reliance 6U3ZR	Duple Dominant II	C55F	1978	Ex Rigby, Patricroft, 1993

Previous Registrations:

TDK291S WND246S, RWB791

Livery: White and red

Reliance Travel operate in the Tameside District, including the Denton service 345 which passes through Dukinfield and Audenshaw. YNA358M, a Daimler Fleetline originating with SELNEC is now back home after a spell with East Yorkshire in Hull. It is seen on one of the housing estates in Denton.
Ken Grime

RIBBLE

Ribble Buses Ltd, Frenchwood Avenue, Preston, Lancashire, PR1 4LU.
Stagecoach Manchester Ltd, Frenchwood Avenue, Preston, Lancashire, PR1 4LU.

Depots: George Street, Blackburn; Goodwin Street, Bolton; Bredbury, (Stagecoach Manchester); Eaves Lane, Chorley; Pimlico Road, Clitheroe; Sidings Road, Fleetwood; Owen Road, Lancaster; Heysham Road, Morecambe and Selbourne Street, Preston. **Outstations:** Burnley, Garstang and Ingleton.

135	F135SPX	Dennis Javelin 11SDL1914	Duple 300	B63F	1989	Ex Hampshire Bus, 1991
136	F136SPX	Dennis Javelin 11SDL1914	Duple 300	B63F	1989	Ex Hampshire Bus, 1991
137	F137SPX	Dennis Javelin 11SDL1914	Duple 300	B63F	1989	Ex Hampshire Bus, 1991

138-144 Dennis Javelin 11SDA2129 Plaxton Premiére Interurban DP47F 1993

138	L138BFV	**140**	L140BFV	**142**	L142BFV	**143**	L143BFV	**144**	L144BFV
139	L139BFV	**141**	L141BFV						

145-161 Dennis Javelin 11SDA2133 Plaxton Premiére Interurban DP47F 1993

145	L145BFV	**150**	L150BFV	**153**	L153BFV	**156**	L156BFV	**159**	L159CCW
146	L146BFV	**151**	L151BFV	**154**	L154BFV	**157**	L157BFV	**160**	L160CCW
148	L148BFV	**152**	L152BFV	**155**	L155BFV	**158**	L158BFV	**161**	L161CCW
149	L149BFV								

162-168 Dennis Javelin 11SDA2133 Plaxton Premiére Interurban DP47F 1993 Ex Stagecoach South, 1994

162	L101SDY	**164**	L104SDY	**166**	L102SDY	**167**	L105SDY	**168**	L107SDY
163	L103SDY	**165**	L106SDY						

Ribble Buses took delivery of 23 Dennis Javelins in 1993 for use on longer limited stop services. These are fitted with Plaxton Premiére Interurban bodies, a special version of the Premiére 320 designed in conjunction with Stagecoach. A need for further examples has been generated on the Burnley to Manchester corridor and all seven of those allocated to Stagecoach South were moved north in 1994. One of the originals, 139, L139BFV, is lettered for The Lancashire Coastliner service between Morecambe to Blackpool. *Roy Marshall*

237-256

Volvo B6 | Alexander Dash | DP40F | 1993

237	L237CCW	**241**	L242CCK	**245**	L245CCK	**249**	L249CCK	**253**	L253CCK
238	L238CCW	**242**	L242CCK	**246**	L246CCK	**250**	L250CCK	**254**	L254CCK
239	L239CCW	**243**	L243CCK	**247**	L247CCK	**251**	L251CCK	**255**	L255CCK
240	L240CCW	**244**	L244CCK	**248**	L248CCK	**252**	L252CCK	**256**	L256CCK

257-265

Volvo B6 | Alexander Dash | DP40F | 1993 | Ex Fife Scottish, 1994

257	L667MSF	**259**	L659HKS	**261**	L661MSF	**263**	L663MSF	**265**	L665MSF
258	L668MSF	**260**	L660HKS	**262**	L662MSF	**264**	L664MSF		

277-283

Volvo B6 | Alexander Dash | B40F | 1993 | Ex Cumberland 1994

277	L277JAO	**278**	L278JAO	**279**	L279JAO	**281**	L281JAO	**283**	L283JAO

301	CHH214T	Leyland National 10351B/1R	B44F	1978	Ex Cumberland, 1993
302u	NEO833R	Leyland National 11351A/1R	B49F	1978	Ex Cumberland, 1993
303	SNS825W	Leyland National 2 NL116AL11/1R	B52F	1981	Ex Cumberland, 1993
304	SNS828W	Leyland National 2 NL116AL11/1R	B52F	1981	Ex Cumberland, 1993
305	HHH371V	Leyland National 2 NL116L11/1R	B52F	1980	Ex Cumberland, 1993

306-318

Leyland National 10351B/1R | B44F | 1978-79 | Ex Cumberland, 1993

306u	AHH204T	**307**	AHH201T	**311**	AHH206T	**312**	CHH210T	**318**	AHH208T

357	KHH377W	Leyland National 2 NL116L11/1R	B52F	1980	Ex Cumberland, 1993
358	OLS809T	Leyland National 10351B/1R	B44F	1978	Ex Cumberland, 1993
359	KHH375W	Leyland National 2 NL116L11/1R	B52F	1980	Ex Cumberland, 1993
367u	MAO367P	Leyland National 11351/1R	B52F	1976	Ex Cumberland, 1993
368	MAO368P	Leyland National 11351/1R	B52F	1976	Ex Cumberland, 1993
370	HHH370V	Leyland National 2 NL116L11/1R	B52F	1980	Ex Cumberland, 1993
371u	OLS807T	Leyland National 10351B/1R	B44F	1978	Ex Cumberland, 1993
372	HHH372V	Leyland National 2 NL116L11/1R	B52F	1980	Ex Cumberland, 1993
373u	HHH373V	Leyland National 2 NL116L11/1R	B52F	1980	Ex Cumberland, 1993
374	KHH374W	Leyland National 2 NL116L11/1R	B52F	1980	Ex Cumberland, 1993
375	AHH209T	Leyland National 10351B/1R	B44F	1978	Ex Cumberland, 1993
376	KHH376W	Leyland National 2 NL116L11/1R	B52F	1980	Ex Cumberland, 1993
377	WAO397Y	Leyland National 2 NL116HLXB/1R	B52F	1982	Ex Cumberland, 1993
378	KHH378W	Leyland National 2 NL116L11/1R	B52F	1980	Ex Cumberland, 1993
379u	NHH379W	Leyland National 2 NL116AL11/1R	B52F	1981	Ex Cumberland, 1993
380	NHH380W	Leyland National 2 NL116AL11/1R	B52F	1981	Ex Cumberland, 1993
381	NHH381W	Leyland National 2 NL116AL11/1R	B52F	1981	Ex Cumberland, 1993
382	NHH382W	Leyland National 2 NL116AL11/1R	DP52F	1981	Ex Cumberland, 1993
383	RRM383X	Leyland National 2 NL116AL11/1R	DP52F	1982	Ex Cumberland, 1993

384	CHH211T	Leyland National 10351B/1R	B44F	1978	Ex Cumberland, 1993
385	RRM384X	Leyland National 2 NL116AL11/1R	DP52F	1982	Ex Cumberland, 1993

386-394

Leyland National 2 NL116AL11/1R | B52F | 1981/82 | Ex Cumberland, 1993

386	RRM386X	**388**	SNS822W	**390**	SHH390X	**392**	SHH392X	**394**	SHH304X
387	SHH387X	**389**	SHH389X	**391**	SHH391X	**393**	SHH393X		

395	RRM385X	Leyland National 2 NL116AL11/1R	B52F	1982	Ex Cumberland, 1993
396	WAO396Y	Leyland National 2 NL116HLXB/1R	B52F	1982	Ex Cumberland, 1993
397u	SNS831W	Leyland National 2 NL116AL11/1R	B52F	1981	Ex Cumberland, 1993
398	WAO398Y	Leyland National 2 NL116HLXB/1R	B52F	1982	Ex Cumberland, 1993
399	SHH388X	Leyland National 2 NL116AL11/1R	B52F	1982	Ex Cumberland, 1993

401-412

Volvo B10M-55 | Alexander PS | DP48F | 1994

401	L341KCK	**403**	L343KCK	**405**	L345KCK	**410**	L340KCK	**412**	M412RRN
402	L342KCK	**404**	L344KCK	**409**	L339KCK	**411**	M411RRN		

424u	OLS806T	Leyland National 10351B/1R		B44F	1978	Ex Cumberland, 1993
449	K449YCW	Optare MetroRider	Optare	B31F	1992	Ex Lancaster, 1993
450	K450YCW	Optare MetroRider	Optare	B31F	1992	Ex Lancaster, 1993

Stagecoach Manchester operate the principal service between Hazel Grove, Stockport and Manchester in competition with GMS Buses. Stagecoach allocated new Volvo B6s to the route in 1993 to which have been added seven Volvo B10Ms during 1994. All carry Alexander coachwork with high-back seating. Photographed at the Hazel Grove terminus ***(above)*** **is 244, L244CCK, while seen in Manchester centre** ***(below)*** **is 410, L340KCK one of the longer Volvo B10Ms.**
A Hamer/Andrew Jarosz

501-527	Mercedes-Benz L608D	Reeve Burgess	DP19F*	1986	*523-527 are B20F 523/4/6/7 ex Cumberland, 1990-91

501	D501RCK	**507**	D507RCK	**513**	D513RCK	**516**	D516RCK	**524**	D524RCK
502	D502RCK	**508**	D508RCK	**514**	D514RCK	**521**	D521RCK	**526**	D526RCK
505	D505RCK	**510**	D510RCK	**515**	D515RCK	**523**	D523RCK	**527**	D527RCK
506	D506RCK	**512**	D512RCK						

530	D672SHH	Mercedes-Benz 609D	Ribble/Cumbria Commercials	B20F	1986	

533-564	Mercedes-Benz L608D	Reeve Burgess	B20F	1986	533/62 ex Cumberland, 1993/91

533	D33UAO	**540**	D540RCK	**545**	D545RCK	**550**	D550RCK	**555**	D555RCK
535	D535RCK	**541**	D541RCK	**546**	D546RCK	**551**	D551RCK	**556**	D556RCK
536	D536RCK	**542**	D542RCK	**547**	D547RCK	**552**	D552RCK	**562**	D562RCK
537	D537RCK	**543**	D543RCK	**548**	D548RCK	**553**	D553RCK	**563**	D563RCK
538	D538RCK	**544**	D544RCK	**549**	D549RCK	**554**	D554RCK	**564**	D564RCK
539	D539RCK								

565-592	Mercedes-Benz 709D	Alexander AM	B23F*	1990	579/80 ex Magicbus, 1990 *567-572 are DP25F

565	G665PHH	**571**	G571PRM	**577**	G577PRM	**583**	G183PAO	**588**	G188PAO
566	G566PRM	**572**	G572PRM	**578**	G578PRM	**584**	G184PAO	**589**	G189PAO
567	G567PRM	**573**	G573PRM	**579**	G179PAO	**585**	G185PAO	**590**	G190PAO
568	G568PRM	**574**	G574PRM	**580**	G180PAO	**586**	G186PAO	**591**	G191PAO
569	G569PRM	**575**	G575PRM	**581**	G181PAO	**587**	G187PAO	**592**	G192PAO
570	G570PRM	**576**	G576PRM	**582**	G182PAO				

595-608	Mercedes-Benz 709D	Alexander AM	B25F	1993	

595	K115XHG	**598**	K118XHG	**600**	K120XHG	**604**	K124XHG	**607**	L127DRN
596	K116XHG	**599**	L119DRN	**602**	L122DRN	**605**	L125DRN	**608**	L128DRN
597	K117XHG								

610-628	Mercedes-Benz 709D	Alexander AM	B23F	1992-93	

610	K610UFR	**614**	K614UFR	**617**	K617UFR	**620**	K620UFR	**625**	K625UFR
611	K611UFR	**615**	K615UFR	**618**	K618UFR	**621**	K621UFR	**627**	K627UFR
612	K612UFR	**616**	K616UFR	**619**	K619UFR	**624**	K624UFR	**628**	K628UFR
613	K613UFR								

629-637	Mercedes-Benz 709D	Alexander AM	B25F	1993	

629	L629BFV	**631**	L631BFV	**633**	L633BFV	**635**	L635BFV	**637**	K112XHG
630	L630BFV	**632**	L632BFV	**634**	L634BFV	**636**	L636BFV		

645u	WAO645Y	Leyland Tiger TRCTL11/2R	Alexander TE	DP47F	1983	Ex Cumberland, 1991
646	WAO646Y	Leyland Tiger TRCTL11/2R	Alexander TE	DP47F	1983	Ex Cumberland, 1991
686	WFR392V	Leyland National 10351B/1R		B44F	1980	
739	UHG739R	Leyland National 11351A/1R		B49F	1976	
800	NCW800T	Leyland National LN113690/1R		B52F	1978	

801-812	Leyland National 10351B/1R		B44F	1977-79	801/2,6-12 ex Cumberland, 1992-93

801	GCW461S	**805**	TRN805V	**807**	TRN807V	**809**	TRN809V	**812**	TRN812V
802u	TRN802V	**806**	TRN806V	**808**	TRN808V				

815-843	Leyland National 2 NL106L11/1R		B44F	1980	813/4/42 ex Cumberland, 1993

813	YRN813V	**820**	YRN820V	**826**	BCW826V	**832**	DBV832W	**838**	DBV838W
814	YRN814V	**821**w	YRN821V	**827**w	BCW827V	**833**	DBV833W	**839**u	DBV839W
815	YRN815V	**822**	YRN822V	**828**	DBV828W	**834**	DBV834W	**840**	DBV840W
817	YRN817V	**823**	BCW823V	**829**	DBV829W	**835**	DBV835W	**841**	DBV841W
818	YRN818V	**824**	BCW824V	**830**	DBV830W	**836**	DBV836W	**842**	DBV842W
819	YRN819V	**825**	BCW825V	**831**	DBV831W	**837**	DBV837W	**843**	DBV843W

Opposite: **In the early 1980s, Ribble took delivery of the shorter Leyland National in considerable numbers, most of which were to the shorter 10.6metre length. Photographed at Lancaster is 861, LFR861X, The Leyland Atlanteans purchased from Lancaster have all been allocated to Preston, now gaining standard Stagecoach livery. One of three with high-back seating is 1221, BFV221Y.**

Ribble
224 LANCASTER
via Promenade & Bare Lane
861
LFR 861X

X60 Sorry-Not in Service
Ribble Buses
1221
BFV 221Y
AGE
Concern

846-877	Leyland National 2 NL106AL11/1R	B44F	1981	847/56/7/60/1/3/72/3 ex Cumberland, 1993

846	JCK846W	**856**	LFR856X	**860**	LFR860X	**868**	LFR868X	**872**	LFR872X
847	JCK847W	**857**	LFR857X	**861**	LFR861X	**870**	LFR870X	**873**	LFR873X
848	JCK848W	**858**	LFR858X	**863**	LFR863X	**871**	LFR871X	**877**	LFR877X
849w	JCK849W	**859**	LFR859X	**866**	LFR866X				

878-886	Leyland National 2 NL116AL11/1R	B52F	1982	881 ex Cumberland, 1993

878	RHG878X	**880**	RHG880X	**881**	RHG881X	**884**	RHG884X	**886**	RHG886X
879	RHG879X								

888-894	Leyland National 2 NL116HLXB/1R	B52F	1983	892 ex Cumberland, 1993

888	ARN888Y	**890**	ARN890Y	**892**	ARN892Y	**893**	ARN893Y	**894**	ARN894Y
889	ARN889Y	**891**	ARN891Y						

895	CEO720W	Leyland National 2 NL116L11/1R		B49F	1980	Ex Cumberland, 1993
896	CEO721W	Leyland National 2 NL116L11/1R		B49F	1980	Ex Cumberland, 1993
897	CEO722W	Leyland National 2 NL116L11/1R		B49F	1980	Ex Cumberland, 1993
898	CEO723W	Leyland National 2 NL116L11/1R		B49F	1980	Ex Cumberland, 1993
900	B900WRN	Leyland Tiger TRCTL11/1R	Duple Dominant	B49F	1984	
1150	A150LFR	Leyland Tiger TRCTL11/2R	Duple Dominant IV	C51F	1983	
1152	B152WRN	Leyland Tiger TRCTL11/2R	Duple Laser 2	C49F	1985	
1157	927GTA	Leyland Tiger TRCTL11/3R	Duple Laser 2	C53F	1985	
1158	B158WRN	Leyland Tiger TRCTL11/3R	Duple Laser 2	C53F	1985	
1159	LJY145	Volvo B10M-61	Plaxton Paramount 3500 III	C48FT	1987	Ex Cumberland, 1993
1160	YDG616	Volvo B10M-61	Plaxton Paramount 3500 III	C48FT	1987	Ex Cumberland, 1993
1163	IIL3507	Volvo B10M-60	Plaxton Paramount 3500 III	C50F	1989	Ex Cumberland, 1993
1164	G117XRE	Volvo B10M-61	Plaxton Expressliner	C46FT	1990	Ex PMT, 1993
1165	G116XRE	Volvo B10M-61	Plaxton Expressliner	C46FT	1990	Ex PMT, 1993
1166	H150CVU	Volvo B10M-60	Plaxton Expressliner	C46FT	1990	Ex Skyliner, Mossley, 1993
1167	H149CVU	Volvo B10M-60	Plaxton Expressliner	C46FT	1990	Ex Skyliner, Mossley, 1993
1200	TCK200X	Leyland Atlantean AN68D/2R	East Lancashire	H50/36F	1982	Ex Lancaster, 1993
1205	LFV205X	Leyland Atlantean AN68C/2R	East Lancashire	H50/36F	1981	Ex Lancaster, 1993
1206	LFV206X	Leyland Atlantean AN68C/2R	East Lancashire	H50/36F	1981	Ex Lancaster, 1993
1212	TCK212X	Leyland Atlantean AN68D/2R	East Lancashire	H50/36F	1982	Ex Lancaster, 1993
1213	WCK213Y	Leyland Atlantean AN68D/2R	East Lancashire	H50/36F	1982	Ex Lancaster, 1993
1214	A214MCK	Leyland Atlantean AN68D/2R	East Lancashire	H50/36F	1984	Ex Lancaster, 1993
1215	WCK215Y	Leyland Atlantean AN68D/2R	East Lancashire	H50/36F	1982	Ex Lancaster, 1993
1221	BFV221Y	Leyland Atlantean AN68D/2R	East Lancashire	DPH45/32F	1983	Ex Lancaster, 1993
1222	BFV222Y	Leyland Atlantean AN68D/2R	East Lancashire	DPH45/32F	1983	Ex Lancaster, 1993
1223	A223MCK	Leyland Atlantean AN68D/2R	East Lancashire	DPH45/32F	1984	Ex Lancaster, 1993
1450	LEO735Y	Leyland Atlantean AN68D/1R	Northern Counties	H43/32F	1983	Ex Barrow, 1989
1451	LEO736Y	Leyland Atlantean AN68D/1R	Northern Counties	H43/32F	1983	Ex Barrow, 1989

1469-1485	Leyland Atlantean AN68A/1R	Eastern Coach Works	H43/31F	1979-80	Ex Cumberland, 1992-93

1469	TRN469V	**1478**	TRN478V	**1481**	TRN481V	**1484**	TRN484V	**1485**	TRN485V
1476	TRN476V	**1480**	TRN480V	**1482**	TRN482V				

2008w	CBV8S	Bristol VRT/SL3/501(Gardner)	Eastern Coach Works	H43/31F	1977	
2021	CBV21S	Bristol VRT/SL3/501(Gardner)	Eastern Coach Works	H43/31F	1977	
2030	DBV30W	Bristol VRT/SL3/6LXB	Eastern Coach Works	H43/31F	1980	
2034	URF662S	Bristol VRT/SL3/501(Gardner)	Eastern Coach Works	H43/31F	1977	Ex Potteries, 1982
2040	FDV813V	Bristol VRT/SL3/6LXB	Eastern Coach Works	H43/31F	1980	Ex Magicbus, 1990
2041w	LDV857W	Bristol VRT/SL3/6LXB	Eastern Coach Works	H43/31F	1980	Ex United Counties, 1990
2042	RRP858R	Bristol VRT/SL3/501	Eastern Coach Works	H43/31F	1977	Ex United Counties, 1990
2043	FDV817V	Bristol VRT/SL3/6LXB	Eastern Coach Works	H43/31F	1980	Ex Magicbus, 1990
2044	FDV833V	Bristol VRT/SL3/6LXB	Eastern Coach Works	H43/31F	1980	Ex Magicbus, 1990
2045	FDV784V	Bristol VRT/SL3/6LXB	Eastern Coach Works	H43/31F	1980	Ex Magicbus, 1990
2051	LFJ882W	Bristol VRT/SL3/6LXC	Eastern Coach Works	H43/31F	1980	Ex United Counties, 1993
2052	LFJ883W	Bristol VRT/SL3/6LXC	Eastern Coach Works	H43/31F	1980	Ex United Counties, 1993
2053	LFJ858W	Bristol VRT/SL3/6LXB	Eastern Coach Works	H43/31F	1980	Ex United Counties, 1993
2054	LFJ859W	Bristol VRT/SL3/6LXB	Eastern Coach Works	H43/31F	1980	Ex United Counties, 1993
2055	LFJ885W	Bristol VRT/SL3/6LXC	Eastern Coach Works	H43/31F	1980	Ex United Counties, 1993
2056	LFJ866W	Bristol VRT/SL3/6LXB	Eastern Coach Works	H43/31F	1980	Ex United Counties, 1993
2057	LFJ861W	Bristol VRT/SL3/6LXB	Eastern Coach Works	H43/31F	1980	Ex United Counties, 1993
2058	LFJ884W	Bristol VRT/SL3/6LXC	Eastern Coach Works	H43/31F	1980	Ex United Counties, 1993
2075	XRR175S	Bristol VRT/SL3/6LXB	Eastern Coach Works	O43/31F	1980	Ex East Midland, 1994
2076	UWV622S	Bristol VRT/SL3/6LXB	Eastern Coach Works	O43/31F	1980	Ex East Kent, 1994

In the last two years Ribble Buses have been successful with tenders for the operation of National Express services, especially those based in the north west of England. Photographed passing along Elizabeth Street near London Victoria is 1167, H149CVU, one of a pair of National Expressliners previously operated by Skyliner. *Colin Lloyd*

The 1994 season is the first for some time where Ribble have provided the open-top service vehicles in the Morecambe area. Signed as the Morecambe and Lancaster Tour Bus, 2076, UWV622S arrived earlier in the year from East Kent and carries a variant of the Stagecoach livery, with red as the dominant body colour. *Ken Grime*

Bristol VRs are now being phased out of the Ribble fleet as a result of the intake of new buses. The remaining Bristols have been transferred to Clitheroe depot, but 2056 was allocated to Chorley when photographed on Lancashire Schools service 791.
Roy Marshall

2100u DBV100W Leyland Olympian B45.02 Eastern Coach Works H45/33F 1980

2101-2137

Leyland Olympian ONLXB/1R* Eastern Coach Works H45/32F 1981-83 *2124-30 are ONLXBT/1R

2101u	GFR101W	**2108**	JFR8W	**2115**	OFV15X	**2122**	OFV22X	**2128**	VRN828Y
2102	JFR2W	**2109**	JFR9W	**2116**	OFV16X	**2123**	OFV23X	**2129**	VRN829Y
2103	JFR3W	**2110**	JFR10W	**2117**	OFV17X	**2124**	SCK224X	**2130**	VRN830Y
2104	JFR4W	**2111**	JFR11W	**2118**	OFV18X	**2125**	SCK225X	**2131**	DBV131Y
2105	JFR5W	**2112**	JFR12W	**2119**	OFV19X	**2126**	SCK226X	**2132**	DBV132Y
2106	JFR6W	**2113**	JFR13W	**2120**	OFV20X	**2127**	VRN827Y	**2137**	DBV137Y
2107	JFR7W	**2114**	OFV14X	**2121**	OFV21X				

2138-2152

Leyland Olympian ONLXB/1R Eastern Coach Works H45/32F 1984

2138	A138MRN	**2142**	A142MRN	**2143**	A143MRN	**2145**	A145MRN	**2152**	B152TRN

2156-2179

Leyland Olympian ONLXB/1R Eastern Coach Works DPH41/26F 1984-85

2156	A156OFR	**2159**	A159OFR	**2171**	C171ECK	**2173**	C173ECK	**2178**	C178ECK
2157	A157OFR	**2170**	C170ECK	**2172**	C172ECK	**2174**	C174ECK	**2179**	C179ECK
2158	A158OFR								

2180-2189

Leyland Olympian ON2R50G16Z4 Alexander RL DPH51/31F 1989

2180	G180JHG	**2182**	G182JHG	**2184**	G184JHG	**2186**	G186JHG	**2188**	G188JHG
2181	G181JHG	**2183**	G183JHG	**2185**	G185JHG	**2187**	G187JHG	**2189**	G189JHG

2191-2197

Leyland Olympian ON2R50G16Z4 Alexander RL H47/30F 1990

2191	H191WFR	**2193**	H193WFR	**2195**	H195WFR	**2196**	H196WFR	**2197**	H197WFR
2192	H192WFR	**2194**	H194WFR						

2198-2210

Leyland Olympian ON2R56G13Z4 Alexander RL DPH47/27F 1991

2198	J198HFR	**2202**	J202HFR	**2205**	J205HFR	**2207**	J207HFR	**2209**	J209HFR
2199	J199HFR	**2203**	J203HFR	**2206**	J206HFR	**2208**	J208HFR	**2210**	J210HFR
2201	J201HFR	**2204**	J204HFR						

2211-2223

Leyland Olympian ONLXB/1R Alexander RL H45/32F 1984-85 Ex Highland Scottish, 1991

2211	A975OST	**2214**	A979OST	**2217**	B893UAS	**2220**	B896UAS	**2222**	B898UAS
2212	A977OST	**2215**	B891UAS	**2218**	B894UAS	**2221**	B897UAS	**2223**	B899UAS
2213	A978OST	**2216**	B892UAS	**2219**	B895UAS				

Previous Registrations:

927GTA	B157WRN	K449YCW	K300LCT	YDG616	D206LWX
IIL3507	F410DUG	K450YCW	K200LCT	LJY145	D205LWX

Operating Companies:

Stagecoach Manchester: 238/42-56, 401-5/9/10, 553

ROBINSONS

Ribblesdale Coachways Ltd, Park Garage, Great Harwood, Lancashire, BB6 7SP

220	E220KFV	DAF MB230LT615	Van Hool Alizée	C51F	1988
221	E221KFV	DAF MB230LT615	Van Hool Alizée	C51F	1988

222-226 DAF MB230LT615 Van Hool Alizée C51FT 1989

222	F222YHG	**223**	F223YHG	**224**	F224YHG	**225**	F225YHG	**226**	F226YHG

227-231 DAF MB230LT615 Van Hool Alizée C49FT 1990

227	G227NCW	**228**	G228NCW	**229**	G229NCW	**230**	G230NCW	**231**	G231NCW

232	G232NCW	DAF MB230LB615	Van Hool Alizée	C51F	1990
233	G233NCW	DAF MB230LB615	Van Hool Alizée	C51F	1990
234	G234NCW	DAF MB230LB615	Van Hool Alizée	C51F	1990
235	G235NCW	DAF MB230LB615	Van Hool Alizée	C51F	1990

236-247 DAF MB230LT615 Van Hool Alizée C49FT 1991-92

236	H236AFV	**239**	H239AFV	**242**	H242AFV	**244**	J244LFR	**246**	J246LFR
237	H237AFV	**240**	H140AFV	**243**	J243LFR	**245**	J245LFR	**247**	J247LFR
238	H238AFV	**241**	H241AFV						

248	L248LFR	Volvo B10M-62	Jonckheere Deauville	C48FT	1994
249	L249LFR	Volvo B10M-62	Jonckheere Deauville	C48FT	1994
250	L250LFR	Volvo B10M-62	Jonckheere Deauville	C48FT	1994

Livery: Green and black.

Robinsons of Great Harwood have been operating quality tours to Europe for many years, each year adding a few new vehicles to the fleet. After many years with Leyland products, eDAF became the standard chassis choice, all with Van Hool Alizée bodywork. The 1994 delivery has seen a change with three Volvo B10Ms joining the fleet, each with a Jonckheere Deauville body. Typical of the fleet is 247, J247LFR, photographed at South Mymms Services while heading for a tour of the Swiss & Italian Lakes. *Colin Lloyd*

ROSE HILL COACHES

A J Lively, 32 Bowden Lane, Marple, Stockport, Greater Manchester, SK6 6ND

MNU472P	Leyland Leopard PSU3C/4R	Plaxton Supreme III Exp	C53F	1976	Ex Whitegate, Royton, 1992
KRE285P	Leyland National 11351/1R		B52F	1976	Ex PMT, 1990
PEH649R	Bristol VRT/SL3/501	Eastern Coach Works	H43/31F	1977	Ex PMT, 1991
PCA419V	Bristol VRT/SL3/501	Eastern Coach Works	H43/31F	1980	Ex PMT, 1990
BYW362V	Leyland National 10351A/2R		B36D	1979	Ex London Buses, 1990
B223AMH	Leyland Royal Tiger B50	Van Hool Alizée	C49FT	1984	Ex London Pullman, Uxbridge, 1991
D837KWT	Freight Rover Sherpa	Dormobile	B16F	1987	Ex West Riding, 1990
D110NOJ	Freight Rover Sherpa	Carlyle	B18F	1987	Ex Bee Line Buzz, 1992
D178NON	Freight Rover Sherpa	Carlyle	B18F	1987	Ex Bee Line Buzz, 1992
D209NON	Freight Rover Sherpa	Carlyle	B18F	1987	Ex Bee Line Buzz, 1992

Livery: Red (buses); brown and white (coaches)

Rose Hill Coaches' Leyland National KRE285P is still in the livery of PMT, tough without fleet names. Based in Marple, this operator is seen working the service between his base and Stockport.
Roy Marshall

ROSSENDALE

Rossendale Transport Ltd, 35 Bacup Road, Rawtenstall,
Lancashire, BB4 7NG

Depots are located at Rawtenstall and Rochdale.

1	F91CWG	MCW MetroRider MF150/105	MCW	B23F	1988	Ex Bee Line Bus, 1991
2	F92CWG	MCW MetroRider MF150/105	MCW	B23F	1988	Ex Bee Line Bus, 1991
3	F93CWG	MCW MetroRider MF150/105	MCW	B23F	1988	Ex Bee Line Bus, 1991
4	F94CWG	MCW MetroRider MF150/105	MCW	B23F	1988	Ex Bee Line Bus, 1991
5	D21CFL	MCW MetroRider MF150/23	MCW	DP25F	1987	Ex Robinson, Kimbolton, 1992
6	E519YWF	MCW MetroRider MF150/35	MCW	B25F	1987	Ex Smith, Prenton, 1989
7	E674DCU	MCW MetroRider MF150/62	MCW	B25F	1987	Ex Moor-Dale, Newcastle, 1990
8	E248UWR	MCW MetroRider MF150/80	MCW	B23F	1988	Ex Yorkshire Rider, 1993
9	D859LND	Renault-Dodge S56	Northern Counties	B20F	1986	Ex Evans, Tregaron, 1992
10	D901MDB	Renault-Dodge S56	Northern Counties	B20F	1986	Ex GM Buses, 1992
11	L911ECW	Optare MetroRider MR09	Optare	B23F	1993	
12	L912ECW	Optare MetroRider MR09	Optare	B23F	1993	
13	L813KCW	Optare MetroRider MR31	Optare	B25F	1994	
14	L814KCW	Optare MetroRider MR31	Optare	B25F	1994	

15-23 Leyland Atlantean AN68A/1R East Lancashire H43/32F 1977-79

15	PTD415S	**17**	PTD417S	**19**	STE19S	**22**	ABN722V	**23**	ABN723V
16	PTD416S	**18**	STE18S	**21**	ABN721V				

24	DDK24W	Leyland Atlantean AN68B/1R	East Lancashire	H43/32F	1980	
25	DDK25W	Leyland Atlantean AN68B/1R	East Lancashire	H43/32F	1980	
26	DDK26W	Leyland Atlantean AN68B/1R	East Lancashire	H43/32F	1980	
27	SND27X	Leyland Atlantean AN68C/1R	East Lancashire	DPH43/27F	1982	
28	SND28X	Leyland Atlantean AN68C/1R	East Lancashire	H43/32F	1982	
29	M529RHG	Volvo Olympian YN2RV18Z4	Alexander Royale RH	H43/29F	1994	
30	M530RHG	Volvo Olympian YN2RV18Z4	Alexander Royale RH	H43/29F	1994	
31	B101PHC	Leyland Olympian ONLXCT/2R	East Lancashire	H47/35F	1985	Ex Stevensons, 1993
32	B102PHC	Leyland Olympian ONLXCT/2R	East Lancashire	H47/35F	1985	Ex Stevensons, 1993
39	PRJ485R	Daimler Fleetline CRG6LXB	Northern Counties	H43/32F	1976	Ex Greater Manchester, 1987
46	RJA705R	Leyland Atlantean AN68A/1R	Northern Counties	H43/32F	1977	Ex GM Buses, 1990
47	SRJ740R	Leyland Atlantean AN68A/1R	Northern Counties	H43/32F	1977	Ex GM Buses, 1990
48	ONF684R	Leyland Atlantean AN68A/1R	Northern Counties	H43/32F	1976	Ex GM Buses, 1990
49	ONF692R	Leyland Atlantean AN68A/1R	Northern Counties	H43/32F	1977	Ex GM Buses, 1990
50	SND550X	Bristol LHS6L	East Lancashire	B28F	1982	
51	SND551X	Bristol LHS6L	East Lancashire	B28F	1982	

Since the closure of GM Buses depot in Rochdale, Rossendale have developed services in that town, and the corridor through to Manchester. For this expansion several Leyland Leopard coaches have been rebodied with new East Lancashire EL2000 bus bodies. One example, seen in Bury bus station working service 483 to Haslingden is 79, PJI9179. *Paul Wigan*

Rossendale have gathered many examples of the MetroRider into the fleet, the latest being new examples from Optare during 1994. Photographed in Rochdale is 61, A61ARN which was delivered new in 1989. *Paul Wigan*

56	E56KHG	MCW MetroRider MF151/8	MCW	B23F	1988	
57	E57KHG	MCW MetroRider MF151/8	MCW	B23F	1988	
58	E58KHG	MCW MetroRider MF151/8	MCW	B23F	1988	
59	E59KHG	MCW MetroRider MF151/8	MCW	B23F	1988	
60	F60ARN	MCW MetroRider MF150/106	MCW	B23F	1989	
61	F61ARN	MCW MetroRider MF150/106	MCW	B23F	1989	
62	F62ARN	MCW MetroRider MF150/106	MCW	B23F	1989	
63	F63ARN	MCW MetroRider MF150/120	MCW	B25F	1989	
64	F164DET	MCW MetroRider MF150/120	MCW	B25F	1989	Ex Liverbus, 1993
65	F165DET	MCW MetroRider MF150/120	MCW	B25F	1989	Ex Liverbus, 1993
67	F171DET	MCW MetroRider MF150/106	MCW	B25F	1989	Ex Liverbus, 1993
68	F168DET	MCW MetroRider MF150/120	MCW	B25F	1989	Ex Liverbus, 1993
69	F169DET	MCW MetroRider MF150/120	MCW	B25F	1989	Ex Liverbus, 1993
70	PJI9170	Leyland Leopard PSU4D/2R	East Lanoo EL2000(1993)	B47F	1977	Ex National Welsh, 1989
71	PJI9171	Leyland Leopard PSU4D/2R	East Lancs EL2000(1993)	B47F	1977	Ex National Welsh, 1989
72	PJI9172	Leyland Leopard PSU3G/4R	Eastern Coach Works B51	C46F	1983	Ex Border, Burnley, 1991
73	PJI9173	Leyland Leopard PSU3G/4R	Eastern Coach Works B51	C47F	1982	Ex Ambassador Travel, 1990
74	PJI9174	Leyland Leopard PSU3G/4R	Eastern Coach Works B51	C47F	1982	Ex Ambassador Travel, 1990
75	PIJ9175	Leyland Leopard PSU3E/4R	East Lancs EL2000 (1993)	DP49F	1980	Ex Blue Bus, Horwich, 1993
76	PJI9176	Leyland Leopard PSU3E/4R	East Lancs EL2000 (1992)	B51F	1977	Ex South Lancs, St Helens, 1993
77	PJI9177	Leyland Leopard PSU3C/4R	East Lancs EL2000 (1992)	B51F	1977	Ex Ellen Smith, Rochdale, 1991
78	PJI9178	Leyland Leopard PSU3E/4R	East Lancs EL2000 (1992)	B51F	1979	Ex Ribble, 1987
79	PJI9179	Leyland Leopard PSU3E/4R	East Lancs EL2000 (1992)	B51F	1980	Ex Cooper, Stockton Heath, 1980
87	OIB1287	Leyland Tiger TRCTL11/2RH	Plaxton Paramount 3200 E	C53F	1984	Ex Kentish Bus, 1990
88	HIL3188	Leyland Olympian ONTL11/2RH	East Lancashire	CH47/31F	1987	
89	LIW4289	Leyland Tiger TRCTL11/3R	Plaxton Paramount 3200 E	C51F	1984	Ex Kentish Bus, 1990
92	F92XBV	Leyland Tiger TRBTL11/2RP	East Lancashire	DP49F	1989	
93	F93XBV	Leyland Tiger TRBTL11/2RP	East Lancashire	B51F	1989	
94	F94XBV	Leyland Tiger TRBTL11/2RP	East Lancashire	B51F	1989	
95	F95XBV	Leyland Tiger TRBTL11/2RP	East Lancashire	B51F	1989	
96	A196WGE	Leyland Tiger TRCTL11/2R	Duple Dominant	B51F	1984	Ex Hutchison, Overtown, 1988
99	OGE9Y	Leyland Tiger TRCTL11/2R	Duple Dominant	B55F	1983	Ex Hutchison, Overtown, 1988

101	H101VFV	Dennis Dart 9SDL3002	Duple/Carlyle Dartline	B36F	1990	
102	H102VFV	Dennis Dart 9SDL3002	Duple/Carlyle Dartline	B36F	1990	
103	H103VFV	Dennis Dart 9SDL3002	Duple/Carlyle Dartline	B36F	1990	
104	H104CHG	Dennis Dart 9SDL3002	Reeve Burgess Pointer	B35F	1991	
105	H105CHG	Dennis Dart 9SDL3002	Reeve Burgess Pointer	B35F	1991	
118	BAJ118Y	Leyland Tiger TRCTL11/2R	Duple Dominant	B53F	1983	Ex Trimdon MS, 1988
119	BAJ119Y	Leyland Tiger TRCTL11/2R	Duple Dominant	B53F	1983	Ex Trimdon MS, 1988
120	BAJ120Y	Leyland Tiger TRCTL11/2R	Duple Dominant	B53F	1983	Ex Trimdon MS, 1988
128	VDY528T	Leyland Atlantean AN68A/2R	East Lancashire	H47/35F	1978	Ex Hastings Top Line, 1989
129	VDY529T	Leyland Atlantean AN68A/2R	East Lancashire	H47/35F	1978	Ex Hastings Top Line, 1989
130	VDY530T	Leyland Atlantean AN68A/2R	East Lancashire	H47/35F	1978	Ex Hastings Top Line, 1989
131	VDY531T	Leyland Atlantean AN68A/2R	East Lancashire	H47/35F	1978	Ex Hastings Top Line, 1989
136	CJK36V	Leyland Atlantean AN68B/1R	East Lancashire	H43/31F	1980	Ex Hastings Top Line, 1989
137	CJK37V	Leyland Atlantean AN68B/1R	East Lancashire	H43/31F	1980	Ex Hastings Top Line, 1989
154	CWG754V	Leyland Atlantean AN68A/1R	Roe	H45/33F	1979	Ex South Yorkshire's Transport, 1991

188-197	Leyland Atlantean AN68B/1R	Alexander AL	H45/33F	1981	Ex South Yorkshire's Transport, 1991

188	JKW288W	**191**	JKW291W	**194**	JKW294W	**196**	JKW296W	**197**	JKW297W
190	JKW290W	**193**	JKW293W						

Typical of the East Lancashire-bodied Leyland Atlanteans in the fleet is 128, VDY528T. Not typical is that this bus was acquired second-hand from Hastings Top Line. It is seen in rather inclement weather at Accrington bus station. *Paul Wigan*

The latest arrival at Rawtenstall is a Volvo Olympian with Alexander's bodywork to their Royale style . Shortly after delivery the vehicle took part in the Ribble Enthusiasts' Club rally from Blackpool to Southport, where it was photographed. *Nick Coleman*

301	OIW5801	Leyland Tiger TRCTL11/3RZ	Van Hool Alizée	C49FT	1985	Ex Lodge, High Easter, 1994
303	OIB5403	Leyland Tiger TRCTL11/3RZ	Plaxton Paramount 3200 II	C53F	1985	Ex Ellen Smith, Rochdale, 1991
304	OIB5404	Leyland Tiger TRCTL11/3R	Plaxton Paramount 3200	C53F	1984	Ex Ellen Smith, Rochdale, 1991
305	OIB6205	Leyland Tiger TRCTL11/3R	Plaxton Paramount 3200	C53F	1983	Ex Ellen Smith, Rochdale, 1991
307	OIB6207	Leyland Leopard PSU5C/5R	Plaxton Supreme IV	C53F	1980	Ex Ellen Smith, Rochdale, 1991
308	OIB3608	Leyland Leopard PSU3E/4R	Plaxton Supreme IV	C49F	1979	Ex Ellen Smith, Rochdale, 1991
309	CFS264S	Leyland Leopard PSU5B/4R	Plaxton Supreme III	C53F	1978	Ex Ellen Smith, Rochdale, 1991
311	LCB652P	Leyland Leopard PSU5A/4RT	Plaxton Supreme III	C57F	1976	Ex Ellen Smith, Rochdale, 1991
314	NIW6514	Leyland Leopard PSU5C/4R	Plaxton Supreme IV	C57F	1985	Ex Parfitt's, Rhymney Bridge, 1991
315	NIW6515	Leyland Tiger TRCTL11/3R	Plaxton Paramount 3500	C49FT	1984	Ex Wimco, Mitcham, 1991
316	NIW6516	Leyland Tiger TRCTL11/3R	Plaxton Paramount 3500	C49FT	1984	Ex Black Horse, London, 1991
317	NIW6517	Volvo B10M-61	Van Hool Alizée	C49FT	1987	Ex Shearings, 1991
318	NIW6518	Volvo B10M-61	Van Hool Alizée	C53F	1986	Ex Shearings, 1991
319	NIW6519	Volvo B10M-61	Van Hool Alizée	C49FT	1988	Ex Rothwell, Heywood, 1993
320	RJI8720	Volvo B10M-61	Van Hool Alizée	C49FT	1989	Ex Shearings, 1993
321	RJI8721	Volvo B10M-60	Van Hool Alizée	C55F	1989	Ex Bruce, Overtown, 1993
322	RJI8722	Volvo B10M-61	Van Hool Alizée	C53F	1987	Ex Clarkson, South Elmsall, 1993
323	RJI8723	Volvo B10M-61	Van Hool Alizée	C49FT	1989	Ex Limebourne, London, 1994
370	170BHR	Volvo B10M-61	Van Hool Alizée	C49FT	1982	Ex Hardman, Waterfoot, 1992
380	LIB1180	Leyland Tiger TRCTL11/3RZ	Plaxton Paramount 3500 II	C49FT	1985	Ex Ribble, 1988
381	LIB1181	Leyland Leopard PSU3F/5R	Plaxton Supreme IV Exp	C53F	1981	Ex Hill's, Tredegar, 1985
382	AAL468A	Leyland Leopard PSU5D/4R	Plaxton P3200 III(1988)	C53F	1980	Ex Border, Burnley, 1993
383	LIB1183	Leyland Tiger TRCTL11/2R	Plaxton Supreme VI Express	C49F	1982	Ex Hill's, Tredegar, 1986
384	LIB1184	Leyland Tiger TRCTL11/2R	Plaxton Supreme VI Express	C53F	1982	Ex Hill's, Tredegar, 1986
390	D741ALR	MCW MetroRider MF150/6	MCW	C16F	1987	Ex Capital, West Drayton, 1990
392	NSU181	MCW MetroRider MF150/9	MCW	DP25F	1986	Ex Patrick Collection, Birmingham, 1992
393	B183FDM	Volvo B10M-50	East Lancashire	CH45/33F	1985	Ex The Wright Company, Wrexham, 1993

Previous Registrations:

170BHR	STT604X, 942AYA	NIW6519	E329OMG	PJI9173	WEX827X
AAL468A	BUH225V	NSU181	D87EDH	PJI9174	XPW876X
HIL3188	D888YHG	OIB1287	A133EPA	PJI9175	WCK139V
LIB1180	B66YFV	OIB3608	LFS487T	PJI9176	VAJ784S
LIB1181	GBO243W	OIB5403	B887WRJ	PJI9177	UGG369R
LIB1183	NDW149X	OIB5404	A886OND	PJI9178	WCK123V
LIB1184	NDW148X	OIB6205	FJA400Y	PJI9179	GRF268V
LIW4289	A148EPA	OIB6207	UFS690V	RJI8720	F763ENE
NIW6514	KAD356V	OIW5801	C396DML	RJI8721	F348JSU
NIW6515	A128MBA	PJI9170	PHB362R	RJI8722	D561MVR, MIW2422, D709NYG
NIW6516	A133MBA	PJI9171	PHB363R	RJI8723	F887SMU
NIW6517	D537MVR	PJI9172	JNH183Y, PJI9175, WEC504Y		
NIW6518	C532DND				

Livery: Cream, red and green (buses); red and white (coaches); white, red and gold (Ellen Smith)

When Burnley FC were playing at Wembley earlier this year, one of the best photographic spots was Staples Corner at the bottom of the M1. Seen in Ellen Smith livery 393, B183FDM, is a Volvo B10M with East Lancashire bodywork previously with The Wright Company in Wrexham.
Keith Grimes

ROYAL MAIL POST BUS

Royal Mail, NW Divisional Headquarters, 79 Oxford Street, Manchester, M60 1DA

Vehicles are based at Ulverston, Broughton-in-Furness, and Penrith.

0750102	H387SET	Vauxhall Midi	Post Office	M11	1991	Penrith-Martindale/Patterdale
1750049	J153LRN	Leyland DAF 400	Post Office	M10	1991	Ulverston-Grizedale
1750050	J154LRN	Leyland DAF 400	Post Office	M10	1991	Broughton-in-Furness
1750051	J155LRN	Leyland DAF 400	Post Office	M10	1991	Barrow reserve

Livery: Post Office red and yellow

SHAW HADWIN

Shaw Hadwin, Billings Road, Dalton-in-Furness, Cumbria, CA16 6TX

Depots are located at Ambleside, Heysham, Stoneleigh, Silverdale and Dalton.

C11	HIL9311	MAN MT8.136F	G C Smith	C28F	1986	Ex Lancaster, 1993
C13	UIB4751	Leyland Tiger TRCTL11/3RH	Duple Laser 2	C57F	1986	Ex Lancaster, 1993
C15	HIL8915	Leyland Tiger TRCTL11/3RZ	Plaxton Paramount 3500 III	C53F	1987	Ex Lancaster, 1993
C16	HIL8916	Leyland Leopard PSU3G/4R	Eastern Coach Works B51	C38FL	1982	Ex Lancaster, 1993
C17	HIL8917	Leyland Tiger TRCTL11/3RH	Plaxton Paramount 3500	C53F	1984	Ex Lancaster, 1993
C98	IIL3198	Leyland Tiger TRCTL11/3RH	Duple 320	C57F	1986	Ex Lancaster, 1993
C99	B99YRN	Leyland Tiger TRCTL11/3RH	Duple Laser 2	C53F	1985	Ex Lancaster, 1993
M13	B59AOP	Ford Transit 190	Carlyle	B16F	1985	Ex Lancaster, 1993
M15	JIL2045	Mercedes-Benz L608D	PMT Hanbridge	C21F	1985	Ex Lancaster, 1993
M18	JIL2047	Mercedes-Benz L608D	PMT Hanbridge	C21F	1985	Ex Lancaster, 1993
101	BYP985	DAF MB200DKFL600	Jonckheere Jubilee	C51FT	1983	Ex Hutchison, Overtown, 1985
102	SIB6724	DAF MB200DKFL600	Plaxton Paramount 3500 II	C51F	1984	
104	UIA7087	Volvo B58-61	Plaxton Supreme III	C53F	1978	
114	A540BEC	Ford Transit VE6	Made-to-Measure	M16	1983	
115	6137RU	Volvo B58-61	Plaxton Supreme IV	C57F	1978	Ex Glenway, Ropponden, 1983
116	KBZ5748	AEC Reliance 6U3ZR	Plaxton Supreme IV	C57F	1979	
117	6267UA	Volvo B10M-61	Jonckheere Jubilee	C51FT	1983	
118	899CAN	Volvo B10M-61	Jonckheere Jubilee	C51FT	1983	
119	GIL3016	Volvo B58-61	Plaxton Supreme IV	C57F	1980	
122	REC52Y	Ford Transit 190	Dormobile	C16F	1986	
125	F485CKU	Mercedes-Benz 609D	Whittaker-Europa	C24F	1988	
128	GSU368	Volvo B10M-61	Jonckheere Jubilee P90	CH51/9FT	1983	Ex SUT, 1988
129	GSU369	Volvo B10M-61	Jonckheere Jubilee P90	CH51/9FT	1983	Ex SUT, 1988
144	THH288X	Mercedes-Benz L508D	Reeve Burgess	C19F	1982	Ex Seaview, Askham, 1986
145	UWR294	DAF MB200DKFL600	Plaxton Paramount 3200	C49FT	1984	Ex Mountain Goat, Windermere, 1987
148	RIB4323	DAF SB2305DHTD585	Plaxton Paramount 3200 III	C57F	1988	
149	GIJ8319	DAF SB2305DHTD585	Plaxton Paramount 3200 III	C57F	1988	
151	PIB9452	DAF SB2305DHTD585	Duple 320	C57F	1988	
152	6682WY	Volvo B58-56	Plaxton Paramount (1984)	C53F	1977	Ex Goodwin, Stockport, 1988
153	XDO32	Volvo B58-56	Plaxton Paramount (1984)	C53F	1977	Ex Goodwin, Stockport, 1988
154	BIW8184	Volvo B58-61	Duple Dominant II	C57F	1979	Ex Bowers, Bridgend, 1989
155	NRG155M	Leyland Atlantean AN68/1R	Alexander AL	H45/36F	1973	Ex Lonsdale, Heysham, 1993
157	509EBL	DAF MB230DKLB615	Plaxton Paramount 3500 III	C51FT	1992	
158	XDL521	DAF MB230DKLB615	Plaxton Paramount 3500 III	C51FT	1989	
159	UIB4752	Volvo B10M-53	Plaxton Paramount 4000 II	CH55/12FT	1984	Ex Mason, Bo'ness, 1989
160	NRG160M	Leyland Atlantean AN68/1R	Alexander AL	H45/36F	1973	Ex Lonsdale, Heysham, 1993
161	XJF386	DAF SB3000DKV601	Van Hool Alizée	C51FT	1990	
162	YFG333	Volvo B58-61	Jonckheere Bermuda	C49FT	1981	Ex Rennie, Dunfermline, 1990
163	H251ANE	Ford Transit VE6	Deansgate	M14	1990	
164	H493BND	Mercedes-Benz 811D	Made-to-Measure	C26F	1990	
165	H176ANE	Ford Transit VE6	Deansgate	M14	1990	
166	C636DEC	Iveco 35-8	Mellor	M15	1985	Ex Brown, Ambleside, 1990
168	H538SEO	Volvo B10M-60	Plaxton Expressliner	C46FT	1991	
169	H539SEO	Volvo B10M-60	Plaxton Expressliner	C46FT	1991	
171	UCK277	Volvo B10M-61	Jonckheere Jubilee P599	C51FT	1983	Ex Rotherham Travel, 1991

172	5129UA	Volvo B10M-61	Jonckheere Jubilee P599	C48FT	1983	Ex Rotherham Travel, 1991
173	824HAO	Volvo B10M-61	Jonckheere Jubilee P599	C51FT	1983	Ex Rotherham Travel, 1991
174w	HIL5695	Volvo B58-61	Jonckheere Bermuda	C49FT	1981	Ex Red Rose, Clayton-le-Dale, 1991
175	DEO722W	Ford Transit 190	Moseley	M12	1980	Ex Percival, Barrow, 1989
176w	SIB6724	MAN 16-290	Jonckheere Deauville	C51FT	1988	Ex Costelloe, Barnoldswick, 1991
177	PNR723	Volvo B58-61	Jonckheere Bermuda	C49FT	1982	Ex Red Rose, Clayton-le-Dale, 1991
178	RIB5092	Volvo B10M-61	Plaxton Paramount 3500 II	C48FT	1988	Ex Wallace Arnold, 1992
179	RIB5093	Volvo B10M-61	Plaxton Paramount 3500 II	C48FT	1988	Ex Wallace Arnold, 1992
181	J728KBC	Volvo B10M-60	Plaxton Paramount 3500 III	C51FT	1992	
182	LSU939	DAF SB2300DHS585	Duple 340	C55FT	1985	Ex C&H, Hambleton, 1992
183	H882JFA	Mercedes-Benz 811D	PMT Ami	C33F	1991	Ex PMT demonstrator, 1992
184	4150RU	Aüwaerter Neoplan N122/3	Aüwaerter Skyliner	CH57/20CT	1989	Ex Mandale, Greystoke, 1993
185	K615EEO	Volvo B10M-60	Van Hool Alizée	C46FT	1993	
186	525XPD	DAF SB3000DKV601	Van Hool Alizée	C49FT	1990	Ex Price, Yeadon, 1993
187	KBZ5750	Mercedes-Benz L608D	Coachcraft	C21F	1983	Ex Richards, Aberteify, 1993
188	G581PVU	Ford Transit VE6	Ford	M8	1989	Ex Tarmac Construction, 1993
189	C963END	Ford Transit VE6	Dormobile	M16FL	1985	Ex ??, 1993
191	F371MTN	Mercedes-Benz L407D	Reeve Burgess	C15F	1988	Ex Target, Cramlington, 1993

SND83X	Leyland Leopard PSU3B/4R	Duple Dominant II(1981)	C51F	1974	Ex Chester, 1994
OVR409W	Ford Transit 190	Ford	M5	1980	Ex ??, 1994
NJI2389	Volvo B10M-61	Jonckheere Bermuda	C49FT	1980	Ex Mountain Goat, Windermere, 1994
TRB25X	Volvo B58-61	Plaxton Supreme IV	C53F	1981	Ex Mountain Goat, Windermere, 1994
GTU606X	Ford Transit 190	Ford	M8	1981	Ex Salford MB, 1989
A383HFK	Ford Transit 190	Ford	M?L	1983	Ex Dudley MB, 1994
KBZ5749	Volvo B10M-61	Duple Dominant IV	C57F	1983	Ex Browns, Ambleside, 1994
D69ONS	Bedford YMP	Plaxton Paramount 3200 II	C35F	1986	Ex Dunnet, Keiss, 1993
E403WRH	Ford Transit VE6	Ford	M8	1987	Ex ??, 1994
JIB3515	Ford Transit VE6	Ford	M14L	1987	Ex ??, 1994
E119UEC	Renault Master T35D	Taylor & Hudson	M15	1988	Ex Browns, Ambleside, 1994

Previous Registrations:

4150RU	F619CWJ	HIL8915	D120GWS	RIB5092	E901UNW
509EBL	F855YJX	HIL8916	A203OCW	RIB5093	E902UNW
5129UA	A305XHE	HIL8917	A203OCW, SND83X	SIB6723	B70CVH, XJS386
525XPD	F217RJX	HIL9311	C854VRY	SIB6724	F597RNV, 525XPD
6137RU	BHD588V	IIL3198	D98SCW	SND83X	HNE643N
6267UA	A538BEC, XDO32	JIB3515	??	UCK277	A314XHE
6682WY	PYT153R, A184UJD	JIL2045	B926DTU	UIA7087	FEC952T
824HAO	A309XHE	JIL2047	B902DTU	UIB4751	C913XEO
899CAN	A539BEC	KBZ5748	EBM444T, 4150RU, TCW781T	UIB4752	B882BMT, USU646
BIW8184	NYS257V	KBZ5749	A287FEC		
BYP985	NNV611Y	KBZ5750	A345JNR, CEC729, A526BHE	UWR294	A739JAY
GIJ8319	E652EEO	LSU939	C786MVH	XDL521	G250EHD
GIL3016	YEC21W, 6SVK	NJI2389	WNV824W, 715BYC, RTT57W	XDO32	PYT152R, A183UJD, 6267UA
GSU368	A315XHE	PIB9452	E467FEO	XJF386	G951KJX
GSU369	A317XHE	PNR723	ENV827X	YFG333	GJR888W
HIL5695	WNH144W	RIB4323	E651EEO		

Livery: White, cream and gold; white, blue and red (National Express) 118/68/9/79

Shaw Hawin have recently added Browns of Ambleside and Lonsdale coaches to the operation, as well as extending their involvement with National Express. Photographed in Parliament Square is 148, RIB4323, a rear-engined DAF SB2305DHS585 with Plaxton Paramount 3200 coachwork. *Colin Lloyd*

SHEARINGS

Shearings Ltd, Lockett Road, Bryn, Ashton-in-Makerfield,
Greater Manchester, WN4 8DE

Depots: Lockett Road, Bryn; Bayton Road, Exhall; Mill Road, Whitwood, Normanton; Acacia Avenue, Sandfields Estate, Port Talbot and Lamberts Road, Tunbridge Wells.

156-200 Volvo B10M-60 Van Hool Alizée C49FT 1992 *179 is C47FT; 181-200 are C53F

156	H156DVM	**165**	H165DVM	**174**	H174DVM	**183**	H183DVM	**192**	H192DVM
157	H157DVM	**166**	H166DVM	**175**	H175DVM	**184**	H184DVM	**193**	H193DVM
158	H158DVM	**167**	H167DVM	**176**	H176DVM	**185**	H185DVM	**194**	H194DVM
159	H159DVM	**168**	H168DVM	**177**	H177DVM	**186**	H186DVM	**195**	H195DVM
160	H160DVM	**169**	H169DVM	**178**	H178DVM	**187**	H187DVM	**196**	H196DVM
161	H161DVM	**170**	H170DVM	**179**	ESU121	**188**	CAZ2818	**197**	H197DVM
162	H162DVM	**171**	H171DVM	**180**	5140RU	**189**	CAZ2819	**198**	H198DVM
163	H163DVM	**172**	H172DVM	**181**	H181DVM	**190**	CAZ3190	**199**	H199DVM
164	H564DVM	**173**	H173DVM	**182**	H182DVM	**191**	H191DVM	**200**	H501DVM

201-205 Van Hool T815H Van Hool Alizée C53F 1991

201	H201DVM	**202**	H202DVM	**203**	H203DVM	**204**	H204DVM	**205**	205CCH

206-257 Volvo B10M-60 Van Hool Alizée C49FT* 1992 *247-257 are C53F

206	J206NNC	**218**	J218NNC	**228**	J228NNC	**238**	J238NNC	**248**	J248NNC
207	J207NNC	**219**	J219NNC	**229**	J229NNC	**239**	J239NNC	**249**	J249NNC
208	J208NNC	**220**	J220NNC	**231**	J231NNC	**241**	J241NNC	**251**	J251NNC
209	J209NNC	**221**	J221NNC	**232**	J232NNC	**242**	J242NNC	**252**	J252NNC
210	J210NNC	**223**	J223NNC	**233**	J233NNC	**243**	J243NNC	**253**	J253NNC
211	J211NNC	**224**	J224NNC	**234**	J234NNC	**244**	J244NNC	**254**	J254NNC
212	J212NNC	**225**	J225NNC	**235**	J235NNC	**245**	J245NNC	**255**	J255NNC
215	J215NNC	**226**	J226NNC	**236**	J236NNC	**246**	J246NNC	**256**	J256NNC
217	J217NNC	**227**	J227NNC	**237**	J237NNC	**247**	J247NNC	**257**	J257NNC

258-268 Volvo B10M-60 Plaxton Premiére 350 C49FT 1992

258	J258NNC	**261**	J261NNC	**263**	J263NNC	**265**	J265NNC	**267**	J263NNC
259	J259NNC	**262**	J262NNC	**264**	J264NNC	**266**	J266NNC	**268**	J263NNC

269-300 Scania K93CRB Plaxton Premiére 320 C53F 1992

269	J269NNC	**275**	J275NNC	**282**	J282NNC	**288**	J288NNC	**294**	J294NNC
270	J270NNC	**276**	J276NNC	**283**	J283NNC	**289**	J289NNC	**295**	J295NNC
271	J271NNC	**277**	J277NNC	**284**	J284NNC	**290**	J290NNC	**296**	J296NNC
272	J272NNC	**278**	J278NNC	**285**	J285NNC	**291**	J291NNC	**297**	J297NNC
273	J273NNC	**279**	J279NNC	**286**	J286NNC	**292**	J292NNC	**298**	J298NNC
274	J274NNC	**281**	J281NNC	**287**	J287NNC	**293**	J293NNC	**299**	J299NNC

300	SPR35	Scania K113CRB	Plaxton Paramount 3500 III	C49FT	1992
301	SPR124	Volvo B10M-60	Van Hool Alizée	C46FT	1992
302	YTP749	Volvo B10M-60	Van Hool Alizée	C46FT	1992

451-485 Volvo B10M-60 Van Hool Alizée C46FT 1993

451	K451VVR	**458**	K458VVR	**465**	K465VVR	**473**	K473VVR	**480**	K80VVR
452	K452VVR	**459**	K459VVR	**466**	K466VVR	**474**	K474VVR	**481**	K481VVR
453	K453VVR	**460**	K460VVR	**467**	K467VVR	**475**	K475VVR	**482**	K482VVR
454	K454VVR	**461**	K461VVR	**468**	K468VVR	**476**	K476VVR	**483**	K483VVR
455	K455VVR	**462**	K462VVR	**469**	K469VVR	**477**	K477VVR	**484**	K484VVR
456	K456VVR	**463**	K463VVR	**471**	K471VVR	**478**	K478VVR	**485**	K485VVR
457	K457VVR	**464**	K464VVR	**472**	K472VVR	**479**	K479VVR		

The Scania K113CRB is being marketed complete with Irizar Century bodywork and twenty have now entered service in the country. Shearings took an example for the 1994 season, numbered 972, L572HRJ, it is seen here at its base in Bryn. For comparison, two Volvo B10Ms were also evaluated, one with Plaxton Excalibur coachwork while the other was fitted with a Jonckheere Deauville body.
Andrew Jarosz

486-495		Volvo B10M-60		Van Hool Alizée		C53F	1993		
486	K486VVR	**488**	K488VVR	**490**	K490VVR	**492**	K492VVR	**494**	K494VVR
487	K487VVR	**489**	K489VVR	**491**	K491VVR	**493**	K493VVR	**495**	K495VVR
496	K496VVR	Volvo B10M-62		Van Hool Alizée		C53F	1993		
561-569		Scania K113CRB		Van Hool Alizée		C46FT	1994		
561	L561FND	**563**	L563FND	**565**	L565FND	**567**	L567FND	**569**	L569FND
562	L562FND	**564**	L564FND	**566**	L566FND	**568**	L568FND		
570	L570FND	Volvo B10M-62		Plaxton Excalibur		C46FT	1994		
571	L571FVU	Volvo B10M-62		Jonckheere Deauville		C43FT	1994		
572	L572HRJ	Scania K113CRB		Irizar Century		C46FT	1994		
650	FSU117	Volvo B10M-61		Van Hool Alizée(1992)		C53F	1988		
719-728		Leyland Tiger TRCL10/3RZM		Plaxton Paramount 3200 III		C53F	1988		
719	F719ENE	**721**	F721ENE	**723**	F723ENE	**725**	F725ENE	**727**	F727ENE
720	F720ENE	**722**	F722ENE	**724**	F724ENE	**726**	F726ENE	**728**	F728ENE
729-752		Volvo B10M-61		Van Hool Alizée		C53F	1989		
729	YXI3043	**734**	YXI3056	**739**	YXI5503	**744**	YXI7906	**749**	YXI9253
730	YXI2730	**735**	YXI3057	**740**	YXI5860	**745**	YXI8421	**750**	YXI9256
731	YXI3049	**736**	YXI3416	**741**	YXI6327	**746**	YXI7923	**751**	YXI9258
732	YXI2732	**737**	YXI4381	**742**	YXI7380	**747**	YXI8897	**752**	YXI2715
733	YXI3254	**738**	YXI4971	**743**	YXI7381	**748**	YXI9243		
805	G805RNC	Scania K93CRB		Plaxton Paramount 3200 III		C53F	1989		

Becoming more traditional in the Shearings fleet is the Van Hool Alizée-bodied Volvo B10M, with many gaining re-registrations during the past year. Photographed at Scratchwood Services while bound for Tarbert is 876, BAZ8576. *Colin Lloyd*

819-845		Leyland Tiger TR2R62C21Z5/8		Plaxton Paramount 3200 III		C53F	1990		
819	G819RNC	**825**	G825RNC	**831**	G831RNC	**836**	G836RNC	**841**	G841RNC
820	G820RNC	**826**	G826RNC	**832**	G832RNC	**837**	G837RNC	**842**	G842RNC
821	G821RNC	**827**	G827RNC	**833**	G833RNC	**838**	G838RNC	**843**	G843RNC
822	G822RNC	**828**	G828RNC	**834**	G834RNC	**839**	G839RNC	**844**	G844RNC
823	G823RNC	**829**	G829RNC	**835**	G835RNC	**840**	G840RNC	**845**	G845RNC
824	G824RNC	**830**	G830RNC						

846-866		Volvo B10M-60		Van Hool Alizée		C53F	1990	866 is C49FT	
846	BAZ6516	**851**	BAZ6851	**855**	BAZ7055	**859**	BAZ7059	**863**	BAZ7918
847	BAZ6527	**852**	BAZ7052	**856**	BAZ7056	**860**	BAZ7360	**864**	BAZ7914
848	BAZ6528	**853**	BAZ7053	**857**	BAZ7057	**861**	BAZ7901	**865**	BAZ7917
849	BAZ7049	**854**	BAZ7054	**858**	BAZ7058	**862**	BAZ7912	**866**	BAZ7916
850	BAZ6170								

876-880		Volvo B10M-60		Van Hool Alizée		C49FT*	1990		
876	BAZ8576	**877**	BAZ8577	**878**	BAZ8578	**879**	XTW359	**880**	WSV528

881	G881VNA	Scania K93CRB	Plaxton Paramount 3200 III	C53F	1990
882	G882VNA	Scania K93CRB	Plaxton Paramount 3200 III	C53F	1990
884	G884VNA	Scania K93CRB	Plaxton Paramount 3200 III	C53F	1990
885	G885VNA	Scania K93CRB	Plaxton Paramount 3200 III	C53F	1990
886	G886VNA	Scania K113CRB	Plaxton Paramount 3500 III	C53F	1990
887	G887VNA	Scania K113CRB	Plaxton Paramount 3500 III	C53F	1990
889	G889VNA	Scania K113CRB	Plaxton Paramount 3500 III	C53F	1990
891	G891VNA	Mercedes-Benz 0303	Plaxton Paramount 3500 III	C53F	1990
892	G892VNA	Mercedes-Benz 0303	Plaxton Paramount 3500 III	C53F	1990
893	G893VNA	Mercedes-Benz 0303	Plaxton Paramount 3500 III	C53F	1990

Representing the Plaxtons in the Shearings fleet 945, H945DRJ, is one of twenty higher 3500 bodies supplied in 1991. Many of the Plaxton Paramounts in the fleet are to the lower 3200 height.
Colin Lloyd

911-920 — Leyland Tiger TR2R62C21Z5/8 — Plaxton Paramount 3200 III — C53F — 1991

911	H411DVM	**913**	H913DRJ	**915**	H915DRJ	**917**	H917DRJ	**919**	H919DRJ
912	H912DRJ	**914**	H911DRJ	**916**	H916DRJ	**918**	H918DRJ	**920**	H920DRJ

922-930 — Scania K93CRB — Plaxton Paramount 3200 III — C53F — 1991

922	H922DRJ	**923**	H523DRJ	**926**	H926DRJ	**929**	H529DRJ	**930**	H930DRJ

931-940 — Volvo B10M-60 — Plaxton Paramount 3200 III — C53F — 1991

931	H931DRJ	**933**	H933DRJ	**935**	H935DRJ	**937**	H937DRJ	**939**	H939DRJ
932	H932DRJ	**934**	H934DRJ	**936**	H936DRJ	**938**	H938DRJ	**940**	H940DRJ

931-960 — Volvo B10M-60 — Plaxton Paramount 3500 III — C49FT — 1991

941	H941DRJ	**945**	H945DRJ	**949**	H949DRJ	**953**	H953DRJ	**957**	H957DRJ
942	H942DRJ	**946**	H946DRJ	**950**	H950DRJ	**954**	H954DRJ	**958**	H958DRJ
943	H943DRJ	**947**	H947DRJ	**951**	H951DRJ	**955**	H955DRJ	**959**	H959DRJ
944	H544DVM	**948**	H948DRJ	**952**	H952DRJ	**956**	H956DRJ	**960**	H960DRJ

Livery: Miami blue, orange and white

Previous Registrations:

205CCH	H502DVM	BAZ7917	G865RNC	YXI3056	F734ENE
BAZ6170	G850RNC	BAZ7918	G863RNC	YXI3057	F735ENE
BAZ6516	G846RNC	BAZ8576	G876VNA	YXI3254	F733ENE
BAZ6527	G847RNC	BAZ8577	G877VNA	YXI3416	F736ENE
BAZ6528	G848RNC	BAZ8578	G878VNA	YXI4381	F737ENE
BAZ6851	G851RNC	CAZ2818	H188DVM	YXI4971	F738ENE
BAZ7049	G849RNC	CAZ2819	H189DVM	YXI5503	F739ENE
BAZ7052	G852RNC	CAZ3190	H590DVM	YXI5860	F740ENE
BAZ7053	G853RNC	ESU117	E650UNE	YXI6327	F741ENE
BAZ7054	G854RNC	SPR124	G106RNC	YXI7380	F742ENE
BAZ7055	G855RNC	SPR35	From new	YXI7381	F743ENE
BAZ7056	G856RNC	WSV528	G880VNA	YXI7906	F744ENE
BAZ7057	G857RNC	XTW359	G879VNA	YXI7923	F746ENE

The Plaxton Premiére 320 and 350 are both represented in the Shearings fleet, the taller vehicle also featuring air conditioning for European tour work. The 320 models are normally found on the British tours, and here at Scratchwood services 272, J272NNC is seen while heading for a few days in Whitby. *Colin Lloyd*

BAZ7058	G858RNC	YTP749	E105UNE	YXI8421	F745ENE
BAZ7059	G859RNC	YXI2715	F752ENE	YXI8897	F747ENE
BAZ7360	G860RNC	YXI2730	F730ENE	YXI9243	F748ENE
BAZ7901	G861RNC	YXI2732	F732ENE	YXI9253	F749ENE
BAZ7912	G862RNC	YXI3043	F729ENE	YXI9256	F750ENE
BAZ7914	G864RNC	YXI3049	F731ENE	YXI9258	F751ENE
BAZ7916	G866RNC				

On order: 58 Volvo B10M-62, Van Hool Alizée for 1995.

SMC TRAVEL

Cannon Travel Ltd, Unit 13 Blackburne Street, Garston Industrial Estate, Liverpool, L19 8JB

MEK92W	Volvo B58-56	Plaxton Supreme IV	C49FT	1980	Ex Amberline, Speke, 1990
D846CRY	Volkswagen LT55	Optare City Pacer	DP25F	1987	Ex Blue Bus, 1994
D847CRY	Volkswagen LT55	Optare City Pacer	DP25F	1987	Ex Blue Bus, 1994
D850CRY	Volkswagen LT55	Optare City Pacer	DP25F	1987	Ex Blue Bus, 1994
D852CRY	Volkswagen LT55	Optare City Pacer	DP25F	1987	Ex Blue Bus, 1994
D343JUM	Volkswagen LT55	Optare City Pacer	B25F	1987	Ex A1A Travel, Birkenhead, 1994
D378JUM	Volkswagen LT55	Optare City Pacer	B25F	1987	Ex A1A Travel, Birkenhead, 1994
F125JGS	Ford Transit VE6	Chassis Developments	C16F	1988	Ex Arena, Speke, 1991
F811RJF	TAZ Dubrava D3500	TAZ	C49FT	1989	Ex Hamilton, Uxbridge, 1992
G711VRY	TAZ Dubrava D3500	TAZ	C49FT	1990	Ex Hamilton, Uxbridge, 1992

Previous Registrations:
A621XFM A145RVL, SEL392

Livery: White, orange and blue, blue (City Pacers)

SOUTH LANCS TRANSPORT

South Lancashire Transport Co Ltd, Unit B, Beaufort Street, Peasley Cross, St Helens, Merseyside, WA9 3BQ

WND477	Leyland Tiger Cub PSUC1/2	Duple Britannia	C41F	1958	Ex preservation, 1993
JDK922P	Leyland Leopard PSU3C/4R	Plaxton Derwent	B48D	1975	Ex Richardson, Sheffield, 1989
JDK924P	Leyland Leopard PSU4C/4R	Plaxton Derwent	B44F	1975	Ex Richardson, Sheffield, 1989
BFV900R	Leyland Leopard PSU4D/2R	Duple Dominant	B47F	1977	Ex Rossendale, 1993
AFB597V	Bristol LH6L	Eastern Coach Works	B43F	1980	Ex Badgerline, 1991
LPY458W	Leyland Leopard PSU3E/4R	Duple Dominant	B55F	1981	Ex United, 1992
LPY462W	Leyland Leopard PSU3E/4R	Duple Dominant	B55F	1981	Ex United, 1992
49XBF	Leyland Leopard PSU3G/4R	Duple Dominant	B53F	1982	Ex A1 (Dunn), Ardrossan, 1991
D154THG	Leyland Tiger TRBTL11/2RP	East Lancashire	B55F	1986	Ex Lancaster, 1993
F71SJX	Mercedes-Benz 709D	Bence	B24F	1989	Ex non-pcv, 1994
G282HDW	Freight Rover Sherpa	Carlyle Citybus 2	B20F	1990	Ex National Welsh, 1992
J3SLT	Mercedes-Benz 709D	Plaxton Beaver	B23F	1992	
J4SLT	Mercedes-Benz 709D	Plaxton Beaver	B23F	1992	
J6SLT	Mercedes-Benz 709D	Plaxton Beaver	B23F	1992	
J7SLT	Mercedes-Benz 709D	Plaxton Beaver	B23F	1992	
J8SLT	Dennis Dart 9.8SDL3003	Plaxton Pointer	B38F	1992	
J9SLT	Dennis Dart 9.8SDL3003	Plaxton Pointer	B38F	1992	
J10SLT	Mercedes-Benz 811D	Plaxton Beaver	B31F	1991	Ex Mercedes-Benz demonstrator, 1993
K455EDT	Mercedes-Benz 811D	Plaxton Beaver	B31F	1992	Ex Mercedes-Benz demonstrator, 1994
K1SLT	Mercedes-Benz 811D	Plaxton Beaver	B31F	1993	
K2SLT	Mercedes-Benz 811D	Plaxton Beaver	B31F	1993	
L1SLT	Dennis Dart 9SDL3011	Plaxton Pointer	B35F	1993	
L2SLT	Dennis Dart 9SDL3011	Plaxton Pointer	B35F	1993	
L11SLT	Dennis Dart 9.8SDL3025	Plaxton Pointer	B38F	1993	

Previous Registrations:

49XBF	OSJ38X	J4SLT	J61MHF	J7SLT	J58MHF
BFV900R	PHB364R, PJI9172	J6SLT	J59MHF	J10SLT	J470XHL
J3SLT	J62MHF				

Livery: Yellow, blue and cream.

South Lancs have acquired many SLT registrations for their modern fleet. L1SLT is a 9-metre Dennis Dart with Plaxton Pointer bodywork, seen working service 14 complete with Merseytravel board. *Simon Watts*

SOUTH MANCHESTER

South Manchester Transport, c/o Adamson Industrial Estate, Raglan Street, Hyde, Greater Manchester, SK14 2DX

PKA724S	Leyland Atlantean AN68/1R	MCW	H43/32F	1978	Ex Merseybus, 1994
PHF559T	Leyland Atlantean AN68/1R	MCW	H43/32F	1978	Ex Merseybus, 1994
JDK912P	Bristol RESL6L	East Lancashire	DP42F	1975	Ex Rossendale, 1994
JDK914P	Bristol RESL6L	East Lancashire	B45F	1975	Ex Rossendale, 1994
KDW707P	Bristol RESL6L	East Lancashire	B47F	1975	Ex Cambrian Coast, Abergynolwyn, 1994

Livery: Green and yellow

South Manchester is one of the latest entrants into the Greater Manchester bus scene, currently sharing premises with Tame Valley in Hyde. Photographed in Stockport while working to Reddish, is JDK912P, a Bristol RESL with East Lancashire bodywork new to Rossendale. *Andrew Jarosz*

SPRINGFIELD COACHWAYS

W & E Trezise, 24 Douglas Bank Drive, Wigan, Greater Manchester, WN6 7NH

XBF58S	Leyland Leopard PSU3E/4R	Duple Dominant I	C49F	1978	Ex PMT, 1990
MFR420T	Ford R1114	Plaxton Supreme III	C53F	1978	Ex Ericsway, Eccles, 1988
JMB331T	Leyland Leopard PSU3E/4R	Duple Dominant II Express	C49F	1979	Ex PMT, 1990
LAO627W	Leyland Leopard PSU3F/4R	Duple Dominant II Express	C53F	1980	Ex Cumberland, 1993
GHR301W	Leyland Leopard PSU3F/4R	Duple Dominant II Express	C53F	1981	Ex Thamesdown, 1993
C150LWA	Leyland Royal Tiger RT	Plaxton Paramount 3500 II	C49FT	1986	Ex Roe, Stainforth, 1993
RJX318	DAF SB2300DHS585	Duple 340	C53F	1987	Ex Hughes DAF, 1990
F817RJF	TAZ Dubrava D3200	TAZ	C49FT	1989	Ex Thandi, Southall, 1990

Previous Registrations:
RJX318 D624YCX

Livery: Blue and white.

Closer inspection of this picture of Springfield Coachways C150LWA will reveal the Leyland Royal Tiger badge displayed on the front. This vehicle is one of the Royal Tiger running units to be bodied away from the Leyland group. Plaxton Paramount 3500 coachwork is fitted,. *Keith Grimes*

STOTTS

Stotts Tours (Oldham) Ltd, 144 Lees Road, Oldham, Greater Manchester, OL4 1HT

XJA549L	Daimler Fleetline CRG6LXB	Park Royal	H43/32F	1973	Ex Greater Manchester, 1988
YNA281M	Daimler Fleetline CRG6LXB	Northern Counties	H43/32F	1973	Ex Greater Manchester, 1988
YNA282M	Daimler Fleetline CRG6LXB	Northern Counties	H43/32F	1973	Ex Greater Manchester, 1987
YNA284M	Daimler Fleetline CRG6LXB	Northern Counties	H43/32F	1973	Ex Greater Manchester, 1988
YNA287M	Daimler Fleetline CRG6LXB	Northern Counties	H43/32F	1973	Ex Greater Manchester, 1988
YNA300M	Daimler Fleetline CRG6LXB	Northern Counties	H43/32F	1973	Ex Greater Manchester, 1988
YNA302M	Daimler Fleetline CRG6LXB	Northern Counties	H43/32F	1973	Ex Greater Manchester, 1987
YNA303M	Daimler Fleetline CRG6LXB	Northern Counties	H43/32F	1974	Ex Greater Manchester, 1987
YNA345M	Daimler Fleetline CRG6LXB	Northern Counties	H43/32F	1974	Ex R & R, Luton, 1990
YNA361M	Daimler Fleetline CRG6LXB	Northern Counties	H43/32F	1974	Ex Caelloi, Pwllheli, 1990
GNC288N	Daimler Fleetline CRG6LXB	Northern Counties	H43/32F	1974	Ex Grey-Green, London, 1990
BNE741N	Daimler Fleetline CRG6LXB	Northern Counties	H43/32F	1974	Ex Greater Manchester, 1988
HJA119N	Daimler Fleetline CRG6LXB	Northern Counties	H43/32F	1975	Ex Grey-Green, London, 1990
RWU52R	Leyland Fleetline FE30AGR	Roe	H43/33F	1976	Ex Yorkshire Rider, 1990
NOC603R	Leyland Fleetline FE30AGR	Park Royal	H43/33F	1976	Ex West Midlands Travel, 1989
WUM105S	Leyland Fleetline FE30AGR	Roe	H44/33F	1976	Ex Yorkshire Rider, 1991
SDA512S	Leyland Fleetline FE30AGR	MCW	H43/33F	1977	Ex West Midlands Travel, 1989
TET747S	Leyland Fleetline FE30AGR	Roe	H43/33F	1977	Ex South Yorkshire's Transport, 1987
TET746S	Leyland Fleetline FE30AGR	Roe	H43/33F	1977	Ex South Yorkshire's Transport, 1987
WDA975T	Leyland Fleetline FE30AGR	MCW	H43/33F	1978	Ex West Midlands Travel, 1989
XSJ652T	Leyland Fleetline FE30AGR	Northern Counties	H44/31F	1979	Ex Western Scottish, 1992
RWM576T	Ford R1114	Plaxton Supreme IV	C53F	1979	Ex Ayton, Nantwich, 1984
SKB650T	Ford R1114	Plaxton Supreme IV	C53F	1979	Ex Ayton, Nantwich, 1984

Stotts of Oldham are now a well-established bus operator in the Oldham area with double-deck vehicles being added to the fleet from a variety of sources. From Western Scottish, and complete with a traditional Scottish-style destintion box, has come XSJ652T. It is seen passing through Oldham heading for Waterhead. *Roy Marshall*

Clydeside 2000 disposed of several Dennis Dominators during 1993, four joining the Stotts fleet. Photographed in Ashton-under-Lyne is TYS256W as it works a local service to Grotton.
Malc McDonald

CWG705V	Leyland Atlantean AN68A/1RSp	Alexander AL	H45/33F	1979	Ex South Yorkshire's Transport, 1991
CWG726V	Leyland Atlantean AN68A/1RSp	Alexander AL	H45/33F	1980	Ex South Yorkshire's Transport, 1991
CWG727V	Leyland Atlantean AN68A/1RSp	Alexander AL	H45/33F	1980	Ex South Yorkshire's Transport, 1991
ECS875V	Leyland Fleetline FE30AGR	Northern Counties	H44/31F	1979	Ex Clydeside 2000, 1993
JKW292W	Leyland Atlantean AN68B/1R	Alexander AL	H45/33F	1981	Ex South Yorkshire's Transport, 1991
JKW298W	Leyland Atlantean AN68B/1R	Alexander AL	H45/33F	1981	Ex South Yorkshire's Transport, 1991
JKW304W	Leyland Atlantean AN68B/1R	Alexander AL	H45/33F	1981	Ex South Yorkshire's Transport, 1991
TYS256W	Dennis Dominator DD135B	Alexander RL	H45/34F	1981	Ex Clydeside 2000, 1993
TYS257W	Dennis Dominator DD135B	Alexander RL	H45/34F	1981	Ex Clydeside 2000, 1993
TYS258W	Dennis Dominator DD135B	Alexander RL	H45/34F	1981	Ex Clydeside 2000, 1993
TYS259W	Dennis Dominator DD135B	Alexander RL	H45/34F	1981	Ex Clydeside 2000, 1993
AFY182X	Leyland Atlantean AN68B/1R	Willowbrook	H45/33F	1981	Ex Somerbus, Paulton, 1994
AFY183X	Leyland Atlantean AN68B/1R	Willowbrook	H45/33F	1981	Ex Somerbus, Paulton, 1994
AFY184X	Leyland Atlantean AN68B/1R	Willowbrook	H45/33F	1981	Ex Somerbus, Paulton, 1994
AFY186X	Leyland Atlantean AN68B/1R	Willowbrook	H45/33F	1981	Ex Merseybus, 1994
AFY189X	Leyland Atlantean AN68B/1R	Willowbrook	H45/33F	1982	Ex Somerbus, Paulton, 1994
AFY190X	Leyland Atlantean AN68B/1R	Willowbrook	H45/33F	1982	Ex Somerbus, Paulton, 1994
GSB146Y	Dennis Dominator DD163	Alexander RL	H45/34F	1983	Ex Clydeside 2000, 1993
GSB147Y	Dennis Dominator DD163	Alexander RL	H45/34F	1983	Ex Clydeside 2000, 1993
GIB1437	Leyland Tiger TRCTL11/3RZ	Plaxton Paramount 3200 II	C53F	1985	Ex Hill's, Tredegar, 1991
D143RAK	Renault-Dodge S56	Reeve Burgess	B25F	1987	Ex Mainline, 1994
D144RAK	Renault-Dodge S56	Reeve Burgess	B25F	1987	Ex Mainline, 1994
D160RAK	Renault-Dodge S56	Reeve Burgess	B25F	1987	Ex Mainline, 1994
D171PYB	Leyland Tiger TRCTL11/3RZ	Plaxton Paramount 3200 II	C53F	1987	Ex Denslow, Chard, 1990
E172UWF	Renault-Dodge S56	Reeve Burgess	B25F	1987	Ex Mainline, 1994
F995HGE	Volvo B10M-60	Plaxton Paramount 3500 III	C53F	1985	Ex Park's, Hamilton, 1990

Previous Registration: GIB1437 B614CKG

Livery: Cream, red and black

STUARTS

Stuarts Bus & Coach Co Ltd, Rothsesey Garage, Broadway, Dukinfield, Greater Manchester, SK16 4UU

33	SXI9033	Leyland Tiger TRCTL11/3R	Duple Dominant IV	C48FT	1982	Ex Gath, Dewsbury, 1989
34	SXI9034	Leyland Leopard PSU5D/4R	Plaxton Supreme IV	C48FT	1982	Ex Midland Red North, 1989
35	YEL95Y	Leyland Leopard PSU5E/4R	Eastern Coach Works B51	DP57F	1982	Ex Midland Red North, 1989
36	SXI3627	DAF MB200DKFL600	Caetano Algarve	C48FT	1984	Ex Brewers, 1991
37	SXI9037	Leyland Leopard PSU3E/4R	Duple Dominant II	C53F	1979	Ex Reliance, Gt Gonerby, 1990
39	SXI9039	DAF MB200DKFL600	LAG Galaxy	C49FT	1986	Ex Elsey, Gosberton, 1991
40	MIW4840	DAF MB200DKFL600	LAG Galaxy	C49FT	1986	Ex Brewers, 1990
41	MIW4841	Bova EL26-581	Bova Europa	C53F	1981	Ex Enterprise, Darlington, 1991
42	628ELX	Leyland Tiger TRCTL11/3R	Duple Dominant IV	C46F	1983	Ex Walls, Wigan, 1993
101	NRG177M	Leyland Atlantean AN68/1R	Alexander AL	H45/29D	1974	Ex Grampian, 1987
103	ABM801A	Leyland Atlantean AN68/1R	Alexander AL	H45/29D	1974	Ex Grampian, 1987
106	SGA710N	Leyland Atlantean AN68/1R	Alexander AL	H45/31F	1974	Ex London Country NE, 1988
109	LRJ213P	Daimler Fleetline CRG6LXB	Roe	H44/34F	1976	Ex Mayne, Manchester, 1988
111	NOC414R	Leyland Fleetline FE30ALR	MCW	H44/34F	1977	Ex West Midlands Travel, 1988
112	KSA187P	Leyland Atlantean AN68A/1R	Alexander AL	H45/29D	1976	Ex Grampian, 1989
116	WDA680T	Leyland Fleetline FE30AGR	Park Royal	H43/33F	1979	Ex West Midlands, 1990
117	WDA980T	Leyland Fleetline FE30AGR	MCW	H43/33F	1979	Ex West Midlands, 1990
118	WDA981T	Leyland Fleetline FE30AGR	MCW	H43/33F	1979	Ex West Midlands, 1991
119	SDA613S	Leyland Fleetline FE30AGR	Park Royal	H43/33F	1978	Ex West Midlands, 1990
120	D109OWG	Renault-Dodge S56	Reeve Burgess Beaver	B25F	1987	Ex South Yorkshire's Transport, 1991
123	K123TCP	DAF SB220LC550	Ikarus Citi Bus	B48F	1992	
124	K124TCP	DAF SB220LC550	Ikarus Citi Bus	B48F	1992	
126	JKW310W	Leyland Atlantean AN68B/1R	Marshall	H45/29D	1981	Ex A1, Ardrossan, 1994
127	JKW322W	Leyland Atlantean AN68B/1R	Marshall	H45/29D	1981	Ex A1, Ardrossan, 1994
128	JKW335W	Leyland Atlantean AN68B/1R	Marshall	H45/29D	1981	Ex A1, Ardrossan, 1994
129	ULS614X	MCW Metrobus DR102/28	MCW	H45/33F	1982	Ex Bullock, Cheadle, 1994
130	D912VCN	Freight Rover Sherpa	Dormobile	B18F	1987	Ex Northumbra, 1994
125	EWF469V	MCW Metrobus DR102/13	MCW	H46/27D	1980	Ex Stevensons, 1993

Previous Registrations:

628ELX	FWH22Y	SXI3627	A504WTH, 948RJO, A659XWN		
ABM801A	NRG173M	SXI9033	NVH961X	SXI9037	CTM408T
MIW4840	C950GTH	SXI9034	SND298X	SXI9039	B596LJU
MIW4841	RSF10Y				

Livery: White, green, red and yellow.

Stuarts 109, LRJ213P, is a local vehicle previously having been operated by Maynes on a similar radial route east from Manchester. While Maynes are to be found on the Ashton Old Road corridor, Stuarts vehicles may be seen on the Hyde Road services. The photograph shows 109 entering Piccadilly, Manchester from the London Road end. *Malc McDonald*

TAME VALLEY

P Keeley, Adamson Industrial Estate, Raglan Street, Hyde,
Greater Manchester, SK14 2DX

24u	PTF724L	Leyland National 1151/2R/0401		B49F	1973	Ex Wirralbus, Birkenhead, 1991
26	SKF26T	Leyland National 11351A/1R		B49F	1979	Ex Merseybus, 1990
30	GUG130N	Leyland National 11351/1R		B49F	1975	Ex Avon Buses, Moreton, 1992
37	YNA337M	Daimler Fleetline CRG6LXB	Northern Counties	H43/32F	1974	Ex Mybus, Hadfield, 1993
44	L644DNA	Mercedes-Benz 709D	Plaxton Beaver	B27F	1993	
45	L645DNA	Mercedes-Benz 709D	Plaxton Beaver	B27F	1993	
53u	JOV753P	Ailsa AB57	Alexander AV	H44/35F	1976	Ex Wombwell Diesels, 1992
57	XWG657T	Leyland Atlantean AN68A/1R	Roe	H45/31F	1978	Ex South Yorkshire's Transport, 1990
81u	JND981N	Leyland Atlantean AN68A/1R	Northern Counties	H43/32F	1975	Ex ??, 1992
93u	AVT493A	Leyland Titan PD2A/27	Weymann	H38/27F	1963	Ex Happy Days, Woodseaves, 1992
97	WVM897S	Leyland Atlantean AN68A/1R	Park Royal	H43/32F	1978	Ex Lancaster, 1991

Previous Registrations:
AVT493A 917DBO

Livery: Green and cream

Tame Valley's 57, XWG657T, was photographed in Stockport bus station prior to operating the service from there south to Hazel Grove. The operator also works service from Stockport to its local area of Hyde and Denton. *Cliff Beeton*

TIMELINE

Timeline Travel Ltd, 12 Bold Street, Leigh, WN7 1AL

Depots Gas Street, Bolton; Moss Industrial Estate, Lowton, Leigh; Railway Yard, Shifnal and GEC Complex, Trafford Park, Manchester.

1	TNA161J	Leyland Leopard PSU3/3R	Alexander AYS	B53F	1971	Ex Shearings, 1992
3	PFS561M	Leyland Leopard PSU3/3R	Alexander AYS	B53F	1974	Ex Progress, Chorley, 1992
5	GLS275S	Leyland Leopard PSU3D/4R	Alexander AT	DP49F	1978	Ex Cumberland, 1992
6	DRJ897L	Leyland Leopard PSU3/3R	Alexander AYS	B53F	1973	Ex Shearings, 1992
7	GLS267S	Leyland Leopard PSU3D/4R	Alexander AT	DP49F	1978	Ex Cumberland, 1992
12	ANJ311T	Leyland Leopard PSU3E/4R	Plaxton Supreme III Express	C49F	1979	Ex Blackburn, 1992
15	ANJ316T	Leyland Leopard PSU3E/4R	Plaxton Supreme III Express	C49F	1979	Ex Blackburn, 1992
16	JND256V	Leyland Leopard PSU5C/4R	Duple Dominant II	C53F	1980	Ex Ribble, 1992
17	JND261V	Leyland Leopard PSU5C/4R	Duple Dominant II	C53F	1980	Ex Ribble, 1992
18	JND262V	Leyland Leopard PSU5C/4R	Duple Dominant II	C53F	1980	Ex Ribble, 1992
19	MRJ279W	Leyland Leopard PSU5D/4R	Plaxton Supreme IV	C50F	1981	Ex Ribble, 1992
20	NNH187Y	Leyland Leopard PSU5C/4R	Duple Dominant IV	C57F	1983	Ex United Counties, 1993
21	NNH191Y	Leyland Leopard PSU5C/4R	Duple Dominant IV	C57F	1983	Ex United Counties, 1993
22	CNH176X	Leyland Leopard PSU3G/4RT	Eastern Coach Works B51	C49F	1982	Ex United Counties, 1993
23	CNH177X	Leyland Leopard PSU3G/4RT	Eastern Coach Works B51	C49F	1982	Ex United Counties, 1993
24	OIB8606	Leyland Tiger TRCTL11/3R	Duple Dominant IV	C57F	1981	Ex Rossendale, 1994
37	F37FNF	Leyland Tiger TRBL10/3ARZA	Alexander N	B55F	1988	Ex Shearings, 1992
38	F38FNF	Leyland Tiger TRBL10/3ARZA	Alexander N	B55F	1988	Ex Shearings, 1992
39	F39FNF	Leyland Tiger TRBL10/3ARZA	Alexander N	B55F	1988	Ex Shearings, 1992
40	F40FNF	Leyland Tiger TRBL10/3ARZA	Alexander N	B55F	1988	Ex Shearings, 1992

49-72 Leyland Tiger TRBL10/3ARZA — Alexander N — B55F — 1989 — Ex Shearings, 1992

49	F49FNF	**54**	G54RND	**59**	G59RND	**64**	G64RND	**69**	G69RND
50	F50FNF	**55**	G55RND	**60**	G60RND	**65**	G65RND	**70**	G70RND
51	F51FNF	**56**	G56RND	**61**	G61RND	**66**	G66RND	**71**	G71RND
52	F52FNF	**57**	G57RND	**62**	G62RND	**67**	G67RND	**72**	G72RND
53	G53RND	**58**	G58RND	**63**	G63RND	**68**	G68RND		

73-88 Volvo Citybus B10M-50 — Alexander Q — B55F — 1990 — Ex Shearings, 1992

73	H73DVM	**77**	H577DVM	**80**	H580DVM	**83**	H83DVM	**86**	H86DVM
74	H74DVM	**78**	H78DVM	**81**	H81DVM	**84**	H84DVM	**87**	H87DVM
75	H575DVM	**79**	H79DVM	**82**	H82DVM	**85**	H85DVM	**88**	H588DVM
76	H76DVM								

Timeline have recently sold some of the Alexander-bodied Leyland Tigers, though twenty-four and the Volvo B10Ms remain. Due shortly are another dozen Mercedes-Benz minibues and six Volvo B6s, all with Alexander bodywork, the Mercedes-Benz being constructed at the Belfast plant where all AMs are being built this year. Passing through St Helens is 84, H84DVM.
Malc McDonald

92	H522FSB	Mercedes-Benz 709D	Reeve Burgess Beaver	B25F	1990	Ex West Coast, Campbeltown, 1994
94	H564OOK	Mercedes-Benz 709D	Carlyle	B27F	1991	Ex The Wright Company, Wrexham, 1993
100	H119YGG	Mercedes-Benz 811D	Dormobile Routemaker	B33F	1991	Ex Pathfinder, Newark, 1993
101	H129YGG	Mercedes-Benz 811D	Dormobile Routemaker	B33F	1991	Ex Pathfinder, Newark, 1993
102	D142NUS	Mercedes-Benz L608D	Alexander	B21F	1986	Ex Kelvin Scottish, 1992
103	D143NUS	Mercedes-Benz L608D	Alexander	B21F	1986	Ex Kelvin Scottish, 1992
105	F545EJA	Mercedes-Benz 709D	PMT	DP23F	1988	Ex Wall, Sharston, 1993
106	H406BVR	Mercedes-Benz 709D	Carlyle	B27F	1990	Ex Starline, Knutsford, 1992

107-112 Mercedes-Benz 609D — Reeve Burgess — B19F — 1989-90 — Ex Shearings, 1992

107 G107RND **108** G108RND **109** H109DVM **110** H110DVM **112** H112DVM

113	G833RDS	Mercedes-Benz 811D	Reeve Burgess Beaver	B33F	1990	Ex Pathfinder, Newark, 1993
114	G411PGG	Mercedes-Benz 811D	Reeve Burgess Beaver	B33F	1989	Ex Allander, Milngavie, 1993
115	G453SGB	Mercedes-Benz 811D	Reeve Burgess Beaver	B33F	1990	Ex Pathfinder, Newark, 1993
117	G737NNS	Mercedes-Benz 811D	Wright	B29F	1990	Ex The Wright Company, Wrexham, 1993
118	G830RDS	Mercedes-Benz 811D	Reeve Burgess Beaver	B33F	1990	Ex Pathfinder, Newark, 1993
119	G902MNS	Mercedes-Benz 811D	Reeve Burgess Beaver	B33F	1989	Ex Allander, Milngavie, 1993
122	RFR418P	Leyland Atlantean AN68/1R	Eastern Coach Works	H43/31F	1976	Ex Ribble, 1993
124	RFR424P	Leyland Atlantean AN68/1R	Eastern Coach Works	H43/31F	1976	Ex Cumberland, 1992
125	SFV425P	Leyland Atlantean AN68/1R	Eastern Coach Works	H43/31F	1976	Ex Cumberland, 1992
126	SFV426P	Leyland Atlantean AN68/1R	Eastern Coach Works	H43/31F	1976	Ex Cumberland, 1992
127	RFR407P	Leyland Atlantean AN68/1R	Eastern Coach Works	H43/31F	1976	Ex Ribble, 1993
128	RFR408P	Leyland Atlantean AN68/1R	Eastern Coach Works	H43/31F	1976	Ex Ribble, 1993
140	F394DOA	Peugeot-Talbot Pullman	Talbot	B18F	1989	Ex Pathfinder, Newark, 1993
141	L141JVR	Peugeot-Talbot Pullman	TBP	B18F	1994	
142	L142JVR	Peugeot-Talbot Pullman	TBP	B18F	1994	
151	L151DRJ	Mercedes-Benz 709D	Alexander AM	B23F	1993	
152	L152DRJ	Mercedes-Benz 709D	Alexander AM	B23F	1993	
153	L153DRJ	Mercedes-Benz 709D	Alexander AM	B23F	1993	
154	L154DRJ	Mercedes-Benz 709D	Alexander AM	B23F	1993	

156-161 Mercedes-Benz 811D — Alexander AM — B31F — 1994

156 M156LNC **157** M157LNC **158** M158LNC **159** M159LNC **160** M160LNC **161** M161LNC

162-167 Mercedes-Benz 709D — Alexander AM — B23F — 1994

162 M162LNC **163** M163LNC **164** M164LNC **165** M165LNC **166** M166LNC **167** M167LNC

201-206 Volvo B6-50 — Alexander Dash — B38F — 1994

201 M201LNC **202** M202LNC **203** M203LNC **204** M204LNC **205** M205LNC **206** M206LNC

699	F699ENE	Leyland Tiger TRCL10/3RZA	Plaxton Paramount 3200 III	C53F	1988	Ex Shearings, 1992
700	F700ENE	Leyland Tiger TRCL10/3RZA	Plaxton Paramount 3200 III	C53F	1988	Ex Shearings, 1992
703	F703ENE	Leyland Tiger TRCL10/3RZA	Plaxton Paramount 3200 III	C53F	1988	Ex Shearings, 1992

Previous Registrations:

DRJ897L BSG526L, 205CCH, DRJ851L, 5140RU
OIB8606 FTD758W
TNA161J RSD732J, SPR35

Livery: Yellow, cream and white.

TOFTWOOD TRAVEL

F G Lovett, 11 Toftwood Avenue, Rainhill, Merseyside, L35 0PU

HSC166X	Leyland Cub CU435	Duple Dominant	B31F	1981	Ex Heaton Travel, Leigh, 1994
D794KWR	Freight Rover Sherpa	Dormobile	B20F	1987	Ex Nip-On, St Helens, 1993
E843JHG	Freight Rover Sherpa	Elme Orion	C16F	1988	Ex Neill, Arnold, 1990

Livery: Blue and white

TOPP-LINE

Toppings Super Coaches Ltd, 123 High Street, Wavertree, Liverpool, L15 8JS

XVU375M	Leyland National 1051/1R		B41F	1974	Ex Greater Manchester, 1987
JNA593N	Leyland National 10351/1R		B41F	1975	Ex Greater Manchester, 1987
VKU80S	Leyland National 11351A/1R		B49F	1978	Ex Chesterfield, 1993
ANA47T	Leyland Fleetline FE30AGR	Northern Counties	H43/32F	1979	Ex Greater Manchester, 1987
VHF57V	Bedford YMT	Plaxton Supreme IV	C53F	1980	
A85OKR	Talbot Express	Dormobile	M12	1984	
TOP659	Leyland Tiger TRCTL11/3R	Plaxton Paramount 3500	C49FT	1984	
TOP11N	Leyland Tiger TRCTL11/3R	Jonckheere Jubilee P50	CH47/12FT	1984	Ex Leyland Bus, 1988
B595ACV	Talbot Express	Dormobile	M12	1984	Ex Sunset Coaches, Sancreed, 1992
B282FNV	Bedford CF	Mellor	M12	1985	Ex ??, 1992
C578GNY	Bedford CF	Dormobile	M12	1985	Ex ??, 1992
C375OEM	Talbot Express	Dormobile	M12	1985	
C446FOV	Ford Transit VE6	Elme Orion	C16F	1986	Ex Leyland Taxis, 1991
D139LTA	Renault-Dodge S56	Reeve Burgess	B23F	1986	Ex Plymouth, 1993
D154LTA	Renault-Dodge S56	Reeve Burgess	B23F	1986	Ex Plymouth, 1993
D300OKM	Mercedes-Benz 609D	Dormobile	B20F	1987	Ex Sunset Coaches, Sancreed, 1990
D894THF	Iveco 35-8	Elme Orion	B16F	1987	
D895THF	Iveco 35-8	Elme Orion	B16F	1987	
E998AKO	Renault-Dodge S56	Dormobile Routemaker	B25F	1988	Ex Dormobile demonstrator, 1990
E325LHG	Peugeot-Talbot Pullman	Talbot	B22F	1988	Ex Dobson, Lostock Gralam, 1993
E567MAC	Peugeot-Talbot Pullman	Talbot	B22F	1988	Ex Barrow, 1989
E235VUD	Volkswagen LT55	Optare City Pacer	B25F	1987	Ex Lancaster, 1993
TOP11	Auwaerter Neoplan N216H	Auwaerter Jetliner	C49FT	1988	
F361YTJ	Leyland Lynx LX112L10ZR1	Leyland	B51F	1988	
F362YTJ	Leyland Lynx LX112L10ZR1	Leyland	B51F	1988	
F363YTJ	Leyland Lynx LX112L10ZR1	Leyland	B51F	1988	
F364YTJ	Leyland Lynx LX112L10ZR1	Leyland	B51F	1988	
F287YKC	Renault-Dodge S56	Dormobile Routemaker	B21F	1989	
G434ETW	Ford Transit VE6	Dormobile	B16F	1990	Ex ?, 1990
G211PGF	Citroen C25	Devon Conversions	M12	1990	Ex ?, 1992
G136GOL	Iveco Daily 49.10	Carlyle Dailybus 2	B21F	1990	Ex ?, 1992
H562GDB	Peugeot-Talbot Pullman	Talbot	B22F	1991	Ex Dobson, Lostock Gralam, 1994

Previous Registrations:

TOP11	E94VWA	TOP11N	B764UHG	TOP659	A726JTJ

Livery: Blue and orange (buses); purple, green and white (coaches); white (most minibuses)

Representing the Topp-Line fleet is Leyland Lynx F361YTJ, one of four of the type in the fleet. It is seen leaving Hamilton Square station for Woodchurch.
Phillip Stephenson

TOWN FLIERS

J & S Pugh, 75 Eastham Crescent, Clock Face, St Helens, Merseyside, WA9 4ER

GTG634W	Leyland Leopard PSU3E/4R	Duple Dominant II	C53F	1980	Ex Cardiff, 1994
ANK317X	Bedford YMQS	Lex Maxeta	B37F	1982	Ex Hutchison, Overtown, 1991
C51HDT	Dennis Domino SDA1202	Optare	B33F	1986	Ex Stevensons, 1992
C54HDT	Dennis Domino SDA1202	Optare	B33F	1986	Ex Stevensons, 1992
C802CBU	Renault-Dodge S56	Northern Counties	B18F	1986	Ex GM Buses, 1991
C806CBU	Renault-Dodge S56	Northern Counties	B18F	1986	Ex GM Buses, 1991
D844LND	Renault-Dodge S56	Northern Counties	B20F	1986	Ex GM Buses, 1991
D231TBW	Volkswagen LT55	Optare City Pacer	B25F	1987	Ex Lancaster, 1993
E748MSC	Mercedes-Benz 709D	Alexander AM	B25F	1988	Ex Carr, Whitburn, 1994
F202XBV	Freight Rover Sherpa	Carlyle Citybus 2	B20F	1989	Ex ABC, Ainsdale, 1993
F795SFA	Renault-Dodge S56	PMT Bursley	B25F	1988	Ex Pride of the Road, Royston, 1993

Livery: Turquoise and white.

ANX317X is unusual in having a Lex Maxeta body on a shortened Bedford YMQ chassis. Now part of the Town Fliers where it is the largest bus at 27-seats, it was photographed in St Helens where it operates on Merseytravel tendered services. *Phillip Stephenson*

VALE OF MANCHESTER

Vales Coaches (Manchester) Ltd, 49 Broughton Street, Manchester, M8 8AN

BAZ7284	Leyland Leopard PSU3B/4R	Plaxton Elite III Express	C53F	1974	Ex Yorkshire Rider, 1994
PDZ7229	Leyland Leopard PSU3D/4R	Duple Dominant I	C53F	1977	Ex East Midland, 1990
URN12V	Leyland Leopard PSU3E/4R	Duple Dominant II Express	C53F	1979	Ex Lancaster, 1991
FAO927V	Leyland National 106L11/2R		B43F	1979	Ex Jim Stones, Glazebury, 1992
VUP514V	Leyland Leopard PSU3E/4R	Duple Dominant	B55F	1980	Ex United, 1992
SSX599V	Seddon Pennine 7	Alexander AYS	B53F	1980	Ex Stevensons, 1992
SSX600V	Seddon Pennine 7	Alexander AYS	B53F	1980	Ex Stevensons, 1992
YSG651W	Seddon Pennine 7	Alexander AYS	B53F	1982	Ex Stevensons, 1992
JFS978X	Seddon Pennine 7	Alexander AYS	B53F	1982	Ex Stonier, Turnstall, 1992
TPL762X	Leyland Tiger TRCTL11/2R	Plaxton Supreme V Express	C53F	1982	Ex Yorkshire Rider, 1994
CAH880Y	Leyland Leopard PSU3G/4R	Eastern Coach Works B51	DP51F	1982	Ex Yorkshire Rider, 1994
B739YUD	Ford Transit 190	Carlyle	B20F	1985	Ex Yorkshire Rider, 1994
E183CNE	Iveco Daily 49.10	Northern Counties	B22F	1988	Ex GM Buses, 1989
E185CNE	Iveco Daily 49.10	Northern Counties	B22F	1988	Ex GM Buses, 1989
E878LHV	Iveco Daily 49.10	Robin Hood City Nippy	B21F	1987	Ex Nat-West Bank, 1990
E250MHX	Iveco Daily 49.10	Robin Hood City Nippy	B21F	1988	Ex Nat-West Bank, 1990
E458TYG	Iveco Daily 49.10	Robin Hood City Nippy	B19F	1988	Ex Yorkshire Rider, 1994
F358FNB	Iveco Daily 49.10	Northern Counties	B22F	1988	
F359FNB	Iveco Daily 49.10	Northern Counties	B22F	1988	
F360FNB	Iveco Daily 49.10	Northern Counties	B22F	1988	
F623YCW	Iveco Daily 49.10	Northern Counties	DP22F	1989	
F624YCW	Iveco Daily 49.10	Northern Counties	B22F	1989	
G874DSC	Mercedes-Benz 609D	Alexander AM	DP25F	1989	Ex Transfleet Services, Coventry, 1993
G150MMA	Iveco Daily 49.10	Robin Hood City Nippy	B21F	1989	Ex Iveco-Ford demonstrator, 1991

Previous Registrations:

BAZ7284	VLB662M, 26BMR, ANW266M	PDZ7229	RCH15R

Livery: Blue and cream

Vale of Manchester are one of many operators to be found on the famous Wilmslow Road corridor radiating south from Manchester, where service 41 will head. One of the reasons this route is so popular is that it links the University, at the city end, with the halls of residence which are nearer to the Didsbury end. Passing under the Arndale centre is VUP514V, a Leyland Leopard with Duple Dominant bus bodywork.
Malc McDonald

VILLAGE

C & J Hunter, 21 Church Road, Garston, Liverpool, Merseyside, L19 8EA

V1	UBV92L	Leyland Atlantean AN68/1R	East Lancashire	H45/31F	1972	Ex Blackburn, 1988
V2	KMW172P	Daimler Fleetline CRG6LXB	Eastern Coach Works	H43/31F	1976	Ex Thamesdown, 1992
V3	JKB146L	Bedford YRQ	Duple Dominant	C45F	1972	Ex Trent, 1978
V4	RCH286R	Leyland Fleetline FE30AGR	Roe	H40/30F	1976	Ex Blue Bus, Derby, 1994
V5	OKY62R	AEC Reliance 6U3ZR	Plaxton Supreme III	C55F	1977	Ex National Travel East, 1981
V6	KWJ137P	Daimler Fleetline CRG6LX	Roe	H42/29D	1976	Ex Pennine Blue, 1994
V7	NKY142R	Leyland Fleetline FE30ALR	Roe	H44/29D	1977	Ex Pennine Blue, 1994
V8	XRR291S	Leyland Fleetline FE30AGR	Northern Counties	H43/30F	1979	Ex Blue Bus, Derby, 1994
V9	NOC598R	Leyland Fleetline FE30ALR	Park Royal	H43/33F	1977	Ex City Fleet, Bootle, 1993
V10	NOC489R	Leyland Fleetline FE30ALR	MCW	H43/33F	1977	Ex City Fleet, Bootle, 1993
V11	XRR295S	Leyland Fleetline FE30AGR	Northern Counties	H43/30F	1979	Ex Blue Bus, Derby, 1994
V12	XRR293S	Leyland Fleetline FE30AGR	Northern Counties	H43/30F	1979	Ex Blue Bus, Derby, 1994
V13	NOC726R	Leyland Fleetline FE30ALR	MCW	H43/33F	1977	Ex ??, 1994
V14	XRR289S	Leyland Fleetline FE30AGR	Northern Counties	H43/30F	1979	Ex Blue Bus, Derby, 1994
V15	RCH288R	Leyland Fleetline FE30AGR	Roe	H40/30F	1976	Ex Blue Bus, Derby, 1994
V16	RCH283R	Leyland Fleetline FE30AGR	Roe	H40/30F	1976	Ex Blue Bus, Derby, 1994
V17	XRR294S	Leyland Fleetline FE30AGR	Northern Counties	H43/30F	1979	Ex Blue Bus, Derby, 1994
V18	KJD80P	Leyland Fleetline FE30ALR	MCW	H44/29F	1976	Ex ??, 1994
V19	KWJ136P	Daimler Fleetline CRG6LX	Roe	H42/29D	1976	Ex Pennine Blue, 1994
V20	SDA658S	Leyland Fleetline FE30AGR	Park Royal	H43/33F	1978	Ex J C Coaches, Widnes, 1994
V21	RCH285R	Leyland Fleetline FE30AGR	Roe	H40/30F	1976	Ex Blue Bus, Derby, 1994
V22	SDA641S	Leyland Fleetline FE30AGR	Park Royal	H43/33F	1978	Ex J C Coaches, Widnes, 1994
V23	XRR296S	Leyland Fleetline FE30AGR	Northern Counties	H43/30F	1979	Ex Blue Bus, Derby, 1994
G5	HDZ6118	Volvo B10M-61	Duple Dominant IV	C53F	1983	Ex Knights, Ellesmere Port, 1989
G6	HDZ9106	Volvo B10M-61	Duple Dominant IV	C53F	1983	Ex Knights, Ellesmere Port, 1989
G7	885NBH	DAF MB200DKTL600	Caetano Algarve	C53F	1984	Ex Knights, Ellesmere Port, 1989

Previous Registrations:

885NBH	A196FAY	HDZ9106	ENF554Y
HDZ6118	ENF563Y	JKB146L	TCH95L

Livery: Red and cream or grey (buses); gold, red, orange and yellow (coaches).

Village V1, UBV92L, is the only Atlantean in an otherwise Fleetline dominated fleet. New to Blackburn it carries locally built East Lancashire bodywork.
Ralph Stevens

WALL'S

M J Wall, Paston Road, Sharston Industrial Estate, Sharston, Manchester, M22 4TF

LNA367	Bedford OB	Duple Vista	C29F	1950	Ex Maypole, Lathom, 1991
YSV608	Daimler Fleetline CRG6LXB	Northern Counties	H43/32F	1972	Ex Greater Manchester, 1986
YSV607	Daimler Fleetline CRG6LXB	Northern Counties	H43/32F	1973	Ex Greater Manchester, 1986
YSV205	Daimler Fleetline CRG6LXB	Northern Counties	H43/32F	1973	Ex Greater Manchester, 1986
YSV662	Leyland Fleetline FE30AGR	MCW	H43/33F	1975	Ex West Midlands Travel, 1990
BNE740N	Daimler Fleetline CRG6LXB	Northern Counties	H43/32F	1975	Ex Mayne, Manchester, 1992
JBN540N	Daimler Fleetline CRG6LXB	Northern Counties	H43/32F	1974	Ex Hulme Hall, Cheadle Hulme, 1993
GNC294N	Daimler Fleetline CRG6LXB	Northern Counties	H43/32F	1975	Ex Mayne, Manchester, 1992
YSV605	Daimler Fleetline CRG6LXB	Northern Counties	H43/32F	1975	Ex Greater Manchester, 1986
YSV589	Daimler Fleetline CRG6LXB	Northern Counties	H43/32F	1975	Ex Greater Manchester, 1986
YSV426	Daimler Fleetline CRG6LXB	Northern Counties	H43/32F	1975	Ex Greater Manchester, 1986
955WAL	Daimler Fleetline CRG6LXB	Northern Counties	H43/32F	1975	Ex Greater Manchester, 1986
168WAL	Daimler Fleetline CRG6LXB	Northern Counties	H43/32F	1975	Ex Greater Manchester, 1986
452WAL	Daimler Fleetline CRG6LXB	Northern Counties	H43/32F	1975	Ex Greater Manchester, 1986
KON332P	Leyland Fleetline FE30AGR	MCW	H43/33F	1975	Ex West Midlands Travel, 1990
RVB977S	Bristol VRT/SL3/6LXB	Willowbrook	O43/31F	1978	Ex East Kent, 1992
BRC838T	Bristol VRT/SL3/6LXB	Eastern Coach Works	H43/31F	1979	Ex Trent, 1993
BRC840T	Bristol VRT/SL3/6LXB	Eastern Coach Works	H43/31F	1979	Ex Trent, 1993
GGM76W	Bristol VRT/SL3/6LXB	Eastern Coach Works	H43/31F	1981	Ex Partridge, Hadleigh, 1992
GGM109W	Bristol VRT/SL3/6LXB	Eastern Coach Works	H43/31F	1981	Ex Partridge, Hadleigh, 1992
HJB460W	Bristol VRT/SL3/6LXB	Eastern Coach Works	H43/31F	1981	Ex Partridge, Hadleigh, 1992
HJB463W	Bristol VRT/SL3/6LXB	Eastern Coach Works	H43/31F	1981	Ex Partridge, Hadleigh, 1992
F823KRJ	Mercedes-Benz 811D	Robin Hood	DP29F	1989	
F318EWF	DAF SB220LC550	Optare Delta	DP49F	1988	Ex Northern Bus, 1990
G251EHD	DAF SB220LC550	Optare Delta	B49F	1989	
G252EHD	DAF SB220LC550	Optare Delta	B49F	1989	
G253EHD	DAF SB220LC550	Optare Delta	B49F	1989	

After a period of operating Fleetline double-decks on its Wilmslow Road services, Wall's have added seven Bristol VRs to the fleet. BRC838T is seen in Didbury from where it runs north to Manchester city centre, terminating in Picadilly. *A Hamer*

Wall's have continued to purchase the DAF SB220LC550 single deck bus, mostly with Optare Delta bodywork. Photographed outside the Refuge Assurance building in Manchester is F318EWF, an example with high-back seating. *Mark Bailey*

G315YHJ	DAF SB220LC550	Optare Delta	B47F	1989	Ex Harris Bus, West Thurrock, 1990
G316YHJ	DAF SB220LC550	Optare Delta	B47F	1989	Ex Harris Bus, West Thurrock, 1990
G776UYV	Iveco Daily 49.10	Robin Hood City Nippy	B21F	1989	Ex Nat-West Bank, 1990
H534YCX	DAF SB220LC550	Optare Delta	B49F	1990	Ex OK, Bishop Auckland, 1992
H536YCX	DAF SB220LC550	Optare Delta	B49F	1990	Ex OK, Bishop Auckland, 1992
H537YCX	DAF SB220LC550	Optare Delta	B49F	1990	Ex OK, Bishop Auckland, 1992
H538YCX	DAF SB220LC550	Optare Delta	B49F	1990	Ex OK, Bishop Auckland, 1992
H611CVM	Ford Transit VE6	Asquith	M12	1991	
H220TCP	DAF SB220LC550	Ikarus CitiBus	B50F	1991	Ex Pride of the Road, 1992
H82PTG	Mercedes-Benz 811D	Optare StarRider	B33F	1991	Ex Bebb, Llantwit Fardre, 1994
H83PTG	Mercedes-Benz 811D	Optare StarRider	B33F	1991	Ex Bebb, Llantwit Fardre, 1994
H84PTG	Mercedes-Benz 811D	Optare StarRider	B33F	1991	Ex Bebb, Llantwit Fardre, 1994
M15WAL	DAF SB220LC550	Ikarus CitiBus	B48F	1994	

Previous Registrations:

160WAL	KBU907P	LNA367	From new	YSV605	YNA357M
452WAL	KBU910P	YSV205	KBU909P	YSV607	YNA359M
955WAL	KBU906P	YSV426	KBU908P	YSV608	VNB241L
JBN540N	GDB165N, CSU917	YSV589	LJA471P	YSV662	KON363P

Livery: Green and white.

WHITE LADY

Whitehead, 1 Glencarron Close, Hoddlesden, Darwen, BB3 3RF

WL9475	Ford R1014	Duple Dominant II	C24F	1981	Ex ? Health Authority, 1993
E919HHG	Mercedes-Benz 609D	Reeve Burgess	C19F	1987	
F956WCK	Mercedes-Benz 609D	Reeve Burgess Beaver	C23F	1988	
H380XHG	Mercedes-Benz 811D	Reeve Burgess Beaver	C33F	1990	
L76ATA	Mercedes-Benz 609D	Devon Conversions	C15FL	1994	

Previous Registrations:
WL9475 ECW549W

Livery: White

Birch Hall Estate, Darwen is where E919HHG was photographed. Belonging to White Lady it is a Mercedes-Benz 609D which carrries convertion by Reeve Burgess from a van rather than a coachbuillt Beaver supplied on a chassis-cowl.
Ken Grime

WIGAN BUS COMPANY

N Stothers & W R Strangeway, 46-52 Richmond Hill, Pemberton, Wigan,Greater Manchester, WN5 8AA

HSC109T	Leyland National 11351A/1R		B49F	1978	Ex Fife Scottish, 1992
HSC112T	Leyland National 11351A/1R		B49F	1978	Ex Fife Scottish, 1992
H81PTG	Mercedes-Benz 811D	Optare StarRider	B33F	1991	Ex Bebb, Llantwit Fardre, 1994
H85PTG	Mercedes-Benz 811D	Optare StarRider	B33F	1991	Ex Bebb, Llantwit Fardre, 1994
L532EHD	DAF SB220LC550	Ikarus CitiBus	B48F	1994	
L533EHD	DAF SB220LC550	Ikarus CitiBus	B48F	1994	

Livery: Red

WINGATES TOURS

Wingates Tours, 43 Spencers Lane, Melling, Liverpool, Merseyside, L31 1HB

MWW561P	Leyland Leopard PSU3C/4R	Plaxton Supreme III Express	C49F	1976	Ex Aintree Coaches, 1990
MIB4197	Volvo B58-61	Plaxton Supreme III	C50F	1978	Ex Goodwin, Stockport, 1989
HGS688T	Volvo B58-61	Plaxton Supreme IV	C55F	1979	Ex Lattimore, Markyate, 1988
YBN631V	Leyland Leopard PSU3E/4R	Plaxton Supreme IV Express	C51F	1979	Ex Chester, 1992
SCH153X	Leyland Leopard PSU3F/4R	Willowbrook 003	C49F	1982	Ex Trent Buses, 1993
PMK106X	Volvo B10M-61	Plaxton Viewmaster IV	C51F	1982	Ex Freeway, Stevenage, 1987
A546HEF	Volvo B10M-61	Duple Caribbean 2	C49FT	1984	Ex Sunspan, Birkenhead, 1991
WSU891	Volvo B10M-61	Caetano Algarve	C49FT	1986	Ex Goodwin, Stockport, 1989
C227EME	Leyland Royal Tiger RT	Van Hool Alizée	C53F	1986	Ex Broomfield, 1993
TOH2S	Volvo B10M-61	Plaxton Paramount 3500 III	C53F	1987	Ex Park's, Hamilton, 1989
E838EUT	Volvo B10M-61	Duple 340	C49FT	1987	Ex Tain Coaches, 1992
E148AJC	Freight Rover Sherpa	Carlyle Citybus 2	B20F	1987	Ex Crosville Wales, 1992
E221XJA	Iveco Daily 49.10	Robin Hood City Nippy	B25F	1988	Ex ?, 1994
F600GET	Scania K113CRB	Van Hool Alizée	C55F	1989	Ex Greenham, Basildon, 1993
K423THG	Mercedes-Benz 711D	PCW Coachworks	B18F	1993	
K427THG	Mercedes-Benz 709D	PCW Coachworks	B18F	1993	

Previous Registrations:

HGS688T	GGD663T, YKJ196, LAT256	WSU891	C691KDS
MIB4197	BGG157S		
TOH2S	D816SGB		

Livery: Pink and white.

ISBN 1 897990 07 3
Typeset by Bill Potter
Published by *British Bus Publishing*
The Vyne, 16 St Margarets Drive, Wellington,
Telford, Shropshire, TF1 3PH
Fax: 01952 255669

Printed by Graphics & Print Ltd
Unit A13, Stafford Park 15
Telford, Shropshire, TF3 3BB